P9-DFN-090

Mathematical Byways

Recreations in Mathematics

Series Editor
David Singmaster

Ayling
Beeling
&
Ceiling

Mathematical Byways

IN AYLING, BEELING, AND CEILING

Hugh ApSimon

Oxford New York Toronto
OXFORD UNIVERSITY PRESS
1984

Oxford University Press, Walton Street, Oxford OX2 6DP
London New York Toronto
Delhi Bombay Calcutta Madras Karachi
Kuala Lumpur Singapore Hong Kong Tokyo
Nairobi Dar es Salaam Cape Town
Melbourne Auckland
and associated companies in
Beirut Berlin Ibadan Mexico City Nicosia

Oxford is a trade mark of Oxford University Press

Published in the United States
by Oxford University Press, New York

British Library Cataloguing in Publication Data
ApSimon, Hugh
Mathematical byways in Ayling, Beeling, and Ceiling. – (Recreations in mathematics)
1. Mathematical recreations
I. Title II. Series
793.7'4 QA95
ISBN 0–19–853201–6

Library of Congress Cataloging in Publication Data
ApSimon, Hugh.
Mathematical byways in Ayling, Beeling, and Ceiling.
(Recreations in mathematics)
1. Mathematical recreations. I. Title. II. Series.
QA95.A67 1984 793.7'4 83–13512
ISBN 0–19–853201–6

Set by Activity Limited
Printed in Great Britain by The Thetford Press Limited,
Thetford, Norfork.

In affectionate memory of
Theo Chaundy
supervisor and friend

PREFACE

It was, I think, in about 1950 that I started keeping notes of mathematical problems that had interested me. (They came from all sorts of sources: a few from exam questions; lots from friends saying 'Hugh; have you heard this one?'; some from my own head.) For a long time I just set out to get the answer — never mind how: that done, I could tuck it away and forget it; there was always a new problem waiting.

A few — a very few — of these problems and their solutions turned out to be worth publishing as formal papers. The vast majority turned out to be trivial, or turgid, or both: page upon page of notes and diagrams that are no longer of the slightest interest to anybody — even me.

There was an in-between group, to which I have kept returning; remembering the dictum of my Oxford supervisor, Theo Chaundy: 'If a problem is capable of a clear and simple statement, and has a clear and simple answer, then there should be a clear and simple path from the one to the other.'

This book is a collection of such problems: they are the ones to which I have eventually — I hope — been able successfully to apply Theo Chaundy's dictum. Their solutions involve very little formal mathematical knowledge: all (but one) require no more than the techniques known to any properly taught 16 year old. But they also — I think — require some ingenuity in their solution: that 16 year old needs to be numerate as well as properly taught.

Most of them are capable of generalization; most of them suggest further, more difficult, problems to attack; some of them are new.

I hope that you enjoy them.

Frimley, Surrey H. ApS.
1983

I should like to thank my long-suffering wife for her continuing forbearance and understanding during the preparation of this book.

CONTENTS

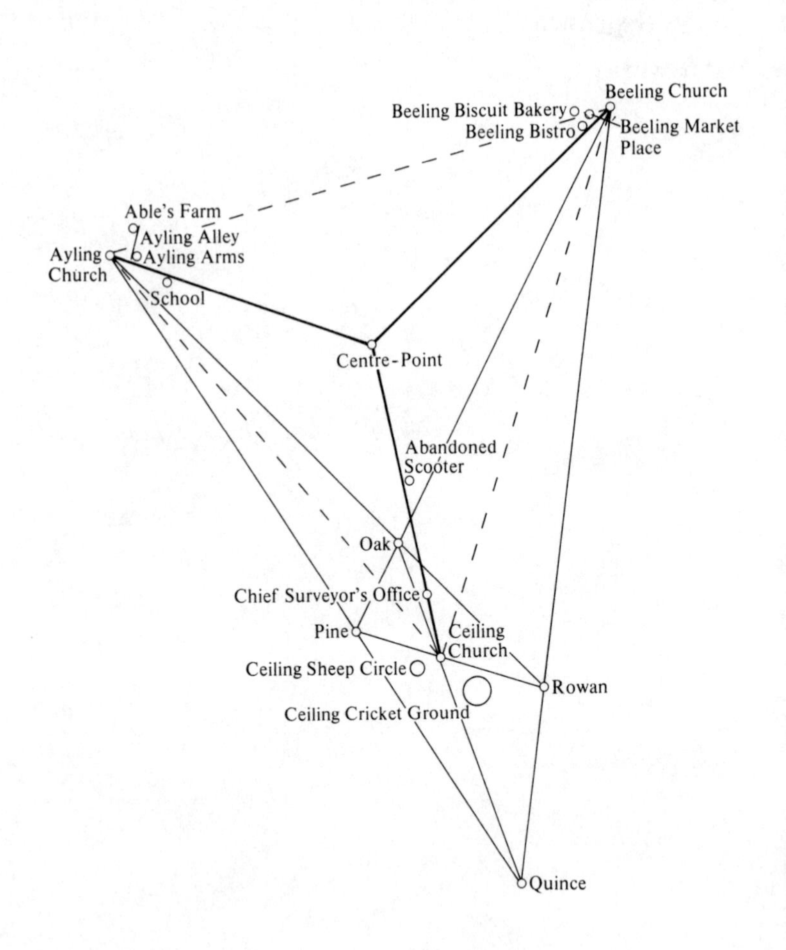

Beeling Church
Beeling Biscuit Bakery
Beeling Bistro
Beeling Market Place
Able's Farm
Ayling Alley
Ayling Arms
Ayling Church
School
Centre-Point
Abandoned Scooter
Oak
Chief Surveyor's Office
Pine
Ceiling Church
Ceiling Sheep Circle
Ceiling Cricket Ground
Rowan
Quince

INTRODUCTION

The general arrangement of each chapter (though I depart from it on occasion) is:

(i) '*Problem*': a specific problem (set in the context of one or all of the villages of Ayling, Beeling, and Ceiling, and involving Tom, or Dick, or Harry, or one or two of their friends and relations), using simple numbers — and probably having an answer that is an equally simple number.

(ii) '*General solution*': an investigation — as simple as I can make it — of the more general problem of which the original problem is a particular example.

(iii) '*Particular solution*': the application of the general solution to the original problem.

(iv) '*Composer's problem*': a general chat about the difficulties I had in setting the problem, or in laying out the solution — or about how I came across the problem in the first place.

(v) '*Extension*': the original problem has triggered an idea for a more difficult one — one that I can't solve: can you?

Right at the end are the answers (to the original specific problems). But they don't really matter very much — unless, of course, you've got the answer right, without looking anywhere except at the problem itself.

If you solve one of the extension problems ... my congratulations!

1

LADDER–BOX

Problem

The prettiest girl in Ayling is Farmer Able's only daughter — and Farmer Able disapproves of Tom as much as Tom and Miss Able approve of each other. Tom has a ladder 18 ft 5 in. long; he has frequently placed it, leaning more steeply than 45°, with its foot on the ground and its top firmly against the wall at Miss Able's window-sill.

Farmer Able recently received some farm equipment in a long packing-case with a square cross-section, 5 ft × 5 ft; he put the packing-case against the wall of his house, immediately under his daughter's window, with one of its long edges fitting flush into the angle between the (horizontal) ground and the (vertical) wall.

Tom was a little bothered when he saw it, but on measuring the cross-section he realized that he could still place his ladder just as he had frequently done before: the ladder would just touch the edge of the packing-case.

How far from the ground is Miss Able's window-sill?

General solution

A ladder of length γ rests squarely against a vertical wall, meeting it at height α; its foot is on the horizontal ground distant β from the base of the wall. We have

$$\alpha^2 + \beta^2 = \gamma^2. \tag{1}$$

A box of square cross-section $\delta \times \delta$ fits flush into the angle of the wall and the ground, and just touches the ladder (see Diagram 1.1). By similar triangles we have

$$\frac{\delta}{\alpha - \delta} = \frac{\beta - \delta}{\delta},$$

so that

$$\frac{1}{\alpha} + \frac{1}{\beta} = \frac{1}{\delta}. \tag{2}$$

There are basically four possible problems that provide two of the values of α, β, γ, δ as data, and then ask the solver to find the value of one (or both) of the other two.

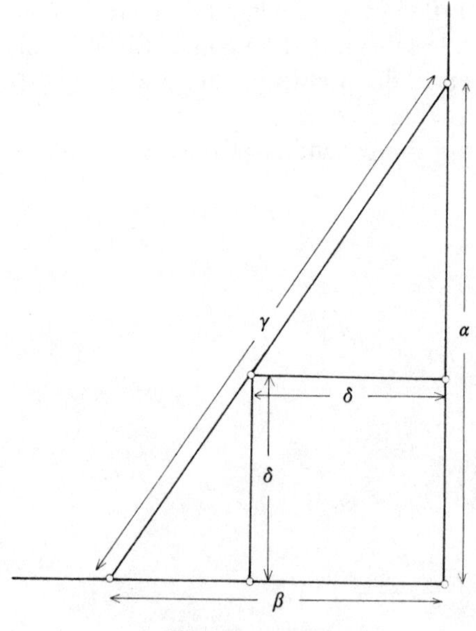

Diagram 1.1

If α, β are given, then γ and δ can be deduced independently of each other, and the problem is trivial.

If α, δ are given, then the deduction of γ is almost as straightforward. Both (1) and (2) are needed, but it does not matter whether the value of β is found explicitly as an intermediate step, or whether β is eliminated between (1) and (2) to give

$$\gamma^2 = \alpha^2 + \frac{\alpha^2\delta^2}{(\alpha - \delta)^2}$$

before the given values of α, δ are inserted.

If α, γ are given, and δ is to be sought, there is a minor pitfall to be avoided. Both (1) and (2) are needed, and if the value of β is found explicitly from (1) before inserting it in (2) all is plain sailing; but if the first step taken is to eliminate β between (1) and (2) the subsequent calculation of δ involves solving a quadratic equation — with at first sight two possible values of δ to choose between.

The most interesting problem occurs when it is the values of γ, δ that are given, and the value of α is sought. There is a temptation to eliminate β between (1) and (2) as a first step; but yielding to it leads to a quartic equation in the unknown α. Starting again, we note that (1) and (2) can be rewritten as

$$(\alpha + \beta)^2 - 2\alpha\beta = \gamma^2, \tag{3}$$

$$\delta(\alpha + \beta) = \alpha\beta;$$

so that

$$(\alpha + \beta)^2 - 2\delta(\alpha + \beta) - \gamma^2 = 0.$$

Since $(\alpha + \beta)$ must be positive, it follows that

$$\alpha + \beta = \delta + \sqrt{(\gamma^2 + \delta^2)}. \tag{4}$$

Rewriting (1) yet again as

$$(\alpha - \beta)^2 + 2\alpha\beta = \gamma^2,$$

we have from (3)

$$(\alpha - \beta)^2 = 2\gamma^2 - (\alpha + \beta)^2; \tag{5}$$

and so from (4)

$$|\alpha - \beta| = \sqrt{[2\gamma^2 - \{\delta + \sqrt{(\gamma^2 + \delta^2)}\}^2]}. \tag{6}$$

We can then obtain α, β immediately from (4) and (6) — though to determine which is which we need further information from the problem-composer (such as that 'the ladder leans more steeply than 45°').

Particular solution

In the particular problem set, we have†

$$\gamma = 221'', \qquad \delta = 60''.$$

So by (4)

$$\begin{aligned}(\alpha + \beta) &= \{60 + \sqrt{(221^2 + 60^2)}\}'' \\ &= (60 + 229)'' \\ &= 289''.\end{aligned}$$

So by (6)

$$\begin{aligned}|\alpha - \beta| &= \{\sqrt{(2 \times 221^2 - 289^2)}\}'' \\ &= 119''.\end{aligned}$$

So

$$\alpha, \beta = 204'', 85''.$$

Since the ladder leans more steeply than 45°, it follows that the height of Miss Able's window-sill from the ground is 204″ — that is, 17′.

Composer's problem

As usual, the composer wants to set an 'integer' problem — one in which not only the data provided for the solver, but also the answer that he is required to find, are whole numbers.

Examination of the composer's problem here would, however, anticipate the 'Meta-ladder–box' problem, which comes next.

So it is omitted.

† In these metric days, some readers may appreciate the reminder that 1 foot = 12 inches, and that — in this context — the superscripts ′ and ″ stand for 'feet' and 'inches'.

2

META-LADDER–BOX

Problem

The local schoolmaster heard about Tom and his ladder, and decided to set a 'Ladder–box' problem to his pupils. He felt that the problem with which Tom had been faced ('given the length of the ladder and the side of the box, find the height of the window-sill') would be too hard for them, so he set the easier problem:

> *A ladder of length ... rests squarely, and more steeply than 45°, against a (vertical) wall, with its foot on the (horizontal) ground distant ... from the base of the wall. A cubical box fits flush into the angle of the wall and the ground, and just touches the ladder. What is the length of side of the box?*

He filled in the gaps in the problem as set with integer numbers of inches, such that the answer would also be an integer number of inches.

To each of the three forms in the school, however, he gave different values for the length of the ladder (all less than 70 ft) and for the distance of its foot from the base of the wall.

The answers submitted by all the pupils were all the same (cribbers!) and all were correct.

What was the answer to the problem the schoolmaster set?

General solution

We look first at the general form of the problem set by the schoolmaster:

> A ladder of length γ'' rests squarely, and more steeply than 45°, against a (vertical) wall, with its foot on the (horizontal) ground distant β'' from the base of the wall. A cubical box of side δ'' fits flush into the angle of the wall and the ground, and just touches the ladder. What is the value of δ?

We have

$$\alpha^2 + \beta^2 = \gamma^2 \tag{1}$$

and

$$\frac{1}{\alpha} + \frac{1}{\beta} = \frac{1}{\delta}, \tag{2}$$

where α is the height of the ladder's point of contact with the wall.

We are told that β, γ, δ are integers. It follows† from (2) that α is rational, and so from (1) that α is an integer.

Let the highest common factor of the integers α, β be e; so there exist integers a, b, e such that

$$\alpha = ea, \qquad \beta = eb,$$

and‡

$$\text{hcf}(a, b) = 1 \tag{3}$$

Then by (2)

$$\delta = \frac{eab}{a + b}.$$

Since δ is an integer we have (using (3)) that $(a + b)$ divides e; so there exists an integer f such that

$$e = f(a + b).$$

† Note: while 'β, γ, δ integer' implies 'α integer' (and 'α, γ, δ integer' implies 'β integer'), it is *not* the case that 'α, β, δ integer' implies 'γ integer' or that 'α, β, γ integer' implies 'δ integer'.

‡ We use the notation 'hcf(p, q) = r' to mean 'r is the highest common factor of p and q'.

Consequently

$$\left.\begin{aligned} \alpha &= fa(a+b), \\ \beta &= fb(a+b), \\ \delta &= fab. \end{aligned}\right\} \tag{4}$$

Now γ is also an integer. From (1) and the first two equations of (4) we have that $f(a+b)$ divides γ; so there exists an integer c such that

$$\gamma = fc(a+b), \tag{5}$$

and

$$a^2 + b^2 = c^2,$$

where (by (3)) no two of a, b, c have a common factor. It follows† that there exist integers p, q, coprime and of opposite parity‡, such that

$$a, b = p^2 - q^2, 2pq$$

(in some order) and

$$c = p^2 + q^2.$$

Inserting these results in (4), (5), and recalling that the ladder slopes at more than 45°, so that $\alpha > \beta$, we have

$$\left.\begin{aligned} &\left.\begin{aligned} \alpha &= \max \\ \beta &= \min \end{aligned}\right\} \left\{\begin{aligned} &f(p^2-q^2)(p^2-q^2+2pq), \\ &2fpq(p^2-q^2+2pq), \end{aligned}\right. \\ &\gamma = f(p^2+q^2)(p^2-q^2+2pq), \\ &\delta = 2f(p^2-q^2)pq, \end{aligned}\right\} \tag{6}$$

as the parametric solution of the problem set by the schoolmaster (and, incidentally, the solution of the deferred 'Composer's

† This is a classic result. If a, b were both odd, then a^2, b^2 would each leave remainder 1 on division by 4, and so $a^2 + b^2$ would leave remainder 2. But c^2 leaves remainder 1 or 0 (according as c is odd or even). So one of a, b must be even: let it be a. By (3) we then have that b and c are both odd. So, since $a^2 = (c+b)(c-b)$, there exists integers p, q such that $c + b = 2p^2$, $c - b = 2q^2$. Hence $b = p^2 - q^2$, $c = p^2 + q^2$, and $a = 2pq$. Since $\text{hcf}(b, c) = 1$ it follows that $\text{hcf}(p, q) = 1$, and that p, q are of opposite parity.

‡ 'Coprime and of opposite parity': a piece of mathematical jargon, but a useful one. 'Coprime' means 'having no common factor'; 'of opposite parity' means 'one is odd and the other is even, but we may not know which is which'.

problem' of the original Ladder–box problem; as far as that problem is concerned all we need to do to be sure of having an 'integer' solution is to assign integer values to f, p, q in (6) and calculate α, β, γ, δ).

Let us call an integer solution of (1), (2) *primitive* if it is one for which, in (6), $f = 1$ and p, q are coprime and of opposite parity. Then any integer solution of (1), (2) is of the form

$$\alpha = f\alpha_0, \quad \beta = f\beta_0, \quad \gamma = f\gamma_0, \quad \delta = f\delta_0,$$

where $(\alpha_0, \beta_0, \gamma_0, \delta_0)$ is a primitive solution. The primitive solutions, in order of increasing δ_0, are tabulated (for $\delta_0 < 1000$) in Table 2.1.

Table 2.1 Primitive solutions of (1), (2).

α_0	β_0	γ_0	δ_0
28	21	35	12
204	85	221	60
345	184	391	120
744	217	775	168
1960	441	2009	360
861	820	1189	420
1645	564	1739	420
4260	781	4331	660

Particular solution

In the particular problem set we are told that corresponding to some value of the length of side of the box (δ'') there are three different values for the length of the ladder (γ''), and that all three values of γ are less than 840.

From Table 2.1 we have that all possible solutions having $\gamma < 840$ are given by

β	γ	δ	
$21u$	$35u$	$12u$	$(u < 24)$
$85v$	$221v$	$60v$	$(v < 4)$
$184w$	$391w$	$120w$	$(w < 3)$
217	775	168	

where u, v, w are integers. It follows by inspection that the values used by the schoolmaster were

β	γ	δ
210	350	
170	442	120
184	391	

so the common answer to the three problems that he set is 120″ (= 10′).

Composer's problem

We are composing a problem that will depend on the fact that, among the integer solutions (α, β, γ, δ) of the two equations

$$\alpha^2 + \beta^2 = \gamma^2$$

$$\frac{1}{\alpha} + \frac{1}{\beta} = \frac{1}{\gamma}$$

(where, without loss of generality, we can assume that $\alpha > \beta$), there are some that share the same value of α, or β, or γ, or **δ**. Which one should we choose?

The smallest value of α that is shared by two different (α, β, γ, δ) is 1428:

α	β	γ	δ
1428	595	1547	420
	1071	1785	612

The smallest value of β that is shared by two different (α, β, γ, δ) is 441:

α	β	γ	δ
588	441	735	252
1960		2009	360

The smallest value of γ that is shared by two different (α, β, γ, δ) is 5083:

α	β	γ	δ
4692	1955	5083	1380
4485	2392		1560

The smallest value of δ that is shared by two different (α, β, γ, δ) is 60:

α	β	γ	δ
140	105	175	60
204	85	221	

We want the solver to be able to concentrate on the essence of the problem, undistracted by having to manipulate large numbers. So it is δ that we choose as the data-element that is to be common to more than one solution.

The next question is: 'How many should "more than one" be?' My decision to select 'three' was somewhat arbitrary: 'two' seemed uninteresting, and 'four' led to undesirably large values for α, β, γ.

There is, however, a considerable temptation to use 'four': there are four different solutions sharing the common value $\delta = 360$, and in them the four values of γ are 1050, 1173, 1326, 2009. But also there are four different solutions sharing the common value $\delta = 420$, and in them the four values of γ are 1189, 1225, 1547, 1739. So a condition in the problem such as 'all the ladder-lengths are less than 150 ft' would allow the answer $\delta = 420$ but not the answer $\delta = 360$: this could be a small (legitimate?) trap for the solver.

Perhaps I should have used 'four' after all.

3

COMPLETE QUADRILATERAL

Problem

The three villages of Ayling, Beeling, and Ceiling lie in a plain. There are four large trees in the plain: an Oak, a Pine, a Quince,† and a Rowan. It is common knowledge that, viewed from Ayling church, the Oak obscures the Rowan, and the Pine obscures the Quince; that, viewed from Beeling church, the Oak obscures the Pine, and the Rowan obscures the Quince; and that the lines from the Oak to the Quince and from the Pine to the Rowan both pass through Ceiling church.

Tom, Dick, and Harry are surveyors. On a recent job they disagreed about that they were supposed to do. Tom was sure that they had to measure the area of the triangle whose corners were the Oak, Pine, and Quince; so they did: it was 4 square miles. Dick was sure that they had to measure the area of the triangle whose corners were the Oak, Quince, and Rowan; so they did: it was 5 square miles. And Harry was sure that they had to measure the area of the triangle whose corners were the Pine, Quince, and Rowan; so they did: it was 6 square miles.

When they reported back to the Chief Surveyor he told them in no uncertain terms that he wasn't interested in trees; what they had been told to do was to measure the area of the triangle whose corners were the churches of Ayling, Beeling, and Ceiling.

'No problem,' said Tom, Dick, and Harry together: 'it must be ... .'

What?

† Sorry about the Quince — but *you* try finding four trees with short names and consecutive initial letters.

General solution

OPQR is a convex quadrilateral.† RO meets QP in A; PO meets QR in B; QO meets PR in C (Diagram 3.1).

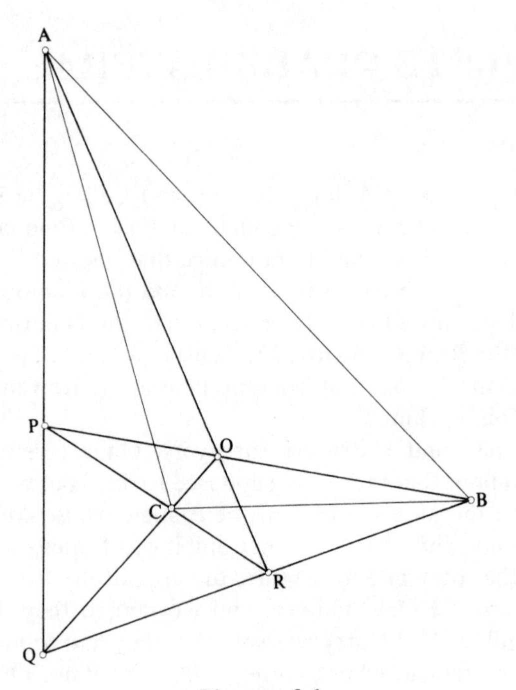

Diagram 3.1

We are given‡ $\triangle OPQ$, $\triangle OPR$, $\triangle PQR$; and we immediately have $\triangle OPR$, since clearly

$$\triangle OPQ + \triangle OQR = \triangle OPR + \triangle PQR. \qquad (1)$$

We want to find $\triangle ABC$.
We need one 'almost obvious' preliminary theorem.

† Strictly speaking (since we are given the *points* O, P, Q, R rather than the *lines* through OP, PQ, QR, RO) OPQR is a quadrangle rather than a quadrilateral. But the term 'quadrilateral' is familiar; 'quadrangle' is less so.

‡ I use '$\triangle XYZ$' to mean 'the area of the triangle XYZ'.

Let DEH and FGH be triangles (sharing the common vertex H) such that D, E, F, G lie in the same straight line (Diagram 3.2).

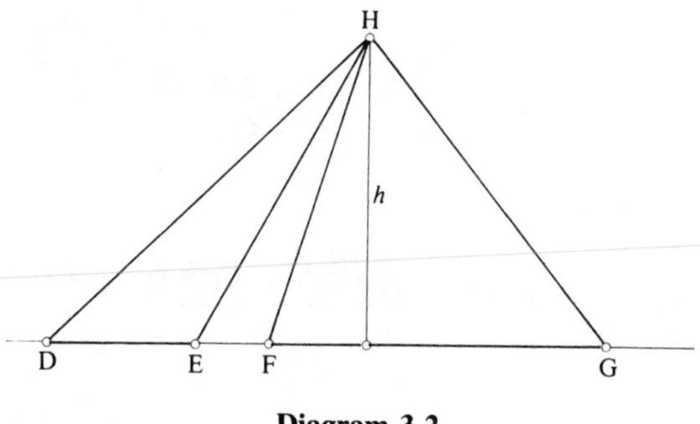

Diagram 3.2

Let the perpendicular distance from H to DEFG be h. The areas of the two triangles are $\frac{1}{2}\text{DE} \times h$, $\frac{1}{2}\text{FG} \times h$ respectively. Hence

$$\frac{\Delta \text{DEH}}{\Delta \text{FGH}} = \frac{\text{DE}}{\text{FG}} . \qquad (2)$$

This is the only theorem that we need in order to solve the problem — but we need to use it a considerable number of times. Think for the moment of A, B, C as 'wild' points (and O, P, Q, R as 'tame' points). Our line of attack will be:

(i) Express $\triangle$ABC (three wild points) in terms of the areas of triangles with just two wild points.
(ii) Then express these in terms of the areas of triangles with at most one wild point.
(iii) Then express these in terms of the areas of triangles with no wild points — that is, in terms of $\triangle$OPQ, $\triangle$OPR, $\triangle$OQR, $\triangle$PQR (which we know).

The first step is easy: clearly

$$\triangle \text{ABC} = \triangle \text{ABO} + \triangle \text{ACO} + \triangle \text{BCO}. \qquad (3)$$

(ABO, ACO, BCO are 'two wild point' triangles.)

The second step involves the multiple use of the theorem (2).

We have, for example,

$$\frac{\Delta\text{ABO}}{\Delta\text{AOP}} = \frac{\text{BO}}{\text{OP}} = \frac{\Delta\text{BOQ}}{\Delta\text{OPQ}},$$

so that

$$\Delta\text{ABO} = \frac{\Delta\text{AOP} \times \Delta\text{BOQ}}{\Delta\text{OPQ}}; \tag{4}$$

and similarly

$$\Delta\text{BCO} = \frac{\Delta\text{BOQ} \times \Delta\text{COR}}{\Delta\text{OQR}}, \tag{5}$$

and

$$\Delta\text{CAO} = \frac{\Delta\text{COR} \times \Delta\text{AOP}}{\Delta\text{OPR}}. \tag{6}$$

(AOP, BOQ, COR are 'one wild point' triangles; OPQ, OPR, OQR are 'tame' triangles.)

The third step again involves the multiple use of the theorem (2). We have, for example,

$$\frac{\Delta\text{AOP}}{\Delta\text{OPR}} = \frac{\text{AO}}{\text{OR}} = \frac{\Delta\text{AOQ}}{\Delta\text{OQR}}$$

$$= \frac{\Delta\text{AOP} + \Delta\text{OPQ}}{\Delta\text{OQR}},$$

so that

$$\Delta\text{AOP} = \frac{\Delta\text{OPQ} \times \Delta\text{OPR}}{\Delta\text{OQR} - \Delta\text{OPR}}; \tag{7}$$

and similarly

$$\Delta\text{BOQ} = \frac{\Delta\text{OPQ} \times \Delta\text{OQR}}{\Delta\text{OPQ} - \Delta\text{OPR}}, \tag{8}$$

and

$$\Delta\text{COR} = \frac{\Delta\text{OPR} \times \Delta\text{OQR}}{\Delta\text{OPQ} + \Delta\text{OQR}}. \tag{9}$$

(All the triangles on the right-hand sides of these last three equations are 'tame'.)

Substituting from (7), (8), (9) in (4), (5), (6), and then in (3), we have (using (1))

$$\Delta ABC = \frac{2 \times \Delta OPQ \times \Delta OPR \times \Delta OQR \times \Delta PQR}{(\Delta OPQ + \Delta OQR)(\Delta OPQ - \Delta OPR)(\Delta OQR - \Delta OPR)} \quad (10)$$

Particular solution

In the particular problem set, we have:

$\Delta OPQ = 4$ square miles,

$\Delta OQR = 5$ square miles,

$\Delta PQR = 6$ square miles;

and so (by (1)) $\Delta OPR = 3$ square miles.

Inserting these values in (10) yields the answer

$\Delta ABC = 40$ square miles.

We could, alternatively, have used the particular values on the way through the working, obtaining successively:

$\Delta AOP = 6$ square miles,

$\Delta BOQ = 20$ square miles,

$\Delta COR = 1\frac{2}{3}$ square miles;

then

$\Delta ABO = 30$ square miles,

$\Delta BCO = 6\frac{2}{3}$ square miles;

$\Delta CAO = 3\frac{1}{3}$ square miles,

and finally

$\Delta ABC = 40$ square miles.

Composer's problem

The main difficulty is that the geometry, though elementary, looks rather forbidding when set out on paper — it certainly did so in my first draft of the General Solution, which occupied about half the space of the present version, and consisted of (in order) equations (7), (8), (9), (4), (5), (6), (3), (10) and very little else except 'hence', 'so that', and 'it follows that'. Logical, but unhelpful. I hope that the present version is more digestible.

Extension

This extension has very little to do with the original problem: it arose as an unexpected side-issue.
Writing for convenience

$$\begin{aligned} \triangle \mathrm{OQR} &= a \text{ square miles}, \\ \triangle \mathrm{OPQ} &= b \text{ square miles}, \\ \triangle \mathrm{OPR} &= c \text{ square miles}, \\ \triangle \mathrm{ABC} &= f \text{ square miles}; \end{aligned}$$

I had from (10)

$$f = \frac{2abc(a + b - c)}{(a + b)(a - c)(b - c)} ;$$

I wanted integer values of a, b, c $(a,b > c > 0)$ that yielded an integer value of f. That was, in itself, no problem. But suppose I had wanted integer values of a, b, c that yielded a *specific* integer value of f: could I have obtained them?

There are certainly some values of f that are 'unrepresentable' in terms of integer a, b, c (the values 1, 2, 3, 4 for example). But $f = 5$ is 'representable' (by $a = 20$, $b = 12$, $c = 2$), and $f = 6$ is 'representable' (by $a = 2$, $b = 2$, $c = 1$).

A number of questions suggest themselves, among them:

1. *Is $f = 7$ representable?*
2. *Is there a largest unrepresentable integer f?*

4

BOWLING AVERAGES

Problem I

Harry and Ken play for the Ceiling Cricket Club, and each is convinced that he's a better bowler than the other. After each match they compare their bowling averages,† and the one with the worse average for the match as a whole buys the drinks.

In the first innings of Ceiling's match against Ayling in the inter-village championship Ken took the first wicket and had 84 runs scored against him; Harry had 252 runs scored against him — but all the same had a better average (for the innings) than Ken did. In the second innings Ken had 252 runs scored against him and Harry had 84 runs scored against him — and again Harry's average (for that innings) was better than Ken's.

But it was Harry who had to buy the drinks after the match.

How many wickets did Ken take in the second innings?

† For non-cricketers: 'bowling average' is 'number of runs scored against' divided by 'number of wickets taken'; the smaller the quotient, the better the average. The total number of wickets taken in an innings cannot exceed 10. There are two innings in a match.

Problem II

Harry and Ken still play for the Ceiling Cricket Club; after each match they compare their bowling averages, and the one with the worse average for the match as a whole buys the drinks.

Ceiling's match against Beeling was similar in some respects to the match against Ayling (but not in all).

In the first innings Harry had 70 runs scored against him; Ken took the first two wickets and had less than 70 runs scored against him — but Harry still had a better average for the innings than Ken did. In the second innings Ken had 70 runs scored against him, and Harry had as many runs scored against him as Ken had had in the first innings; and again Harry's bowling average (for that innings) was better than Ken's.

But it was still Harry who had to buy the drinks after the match. (Harry is getting rather tired of this.)

How many runs did Ken have scored against him in the first innings?

General comment

There are two bowlers, $\mathscr{H}$ and $\mathscr{K}$. Let the number of runs scored against, and the number of wickets taken, be as in Table 4.1.

Table 4.1

	Bowler	$\mathscr{H}$	$\mathscr{K}$
1st innings	Runs against	H_1	K_1
	Wickets	h_1	h_1
	Average	H_1/h_1	K_1/k_1
2nd innings	Runs against	H_2	K_2
	Wickets	h_2	k_2
	Average	H_2/h_2	K_2/k_2
Whole match	Runs against	$H_1 + H_2$	$K_1 + K_2$
	Wickets	$h_1 + h_2$	$k_1 + k_2$
	Average	$\dfrac{H_1 + H_2}{h_1 + h_2}$	$\dfrac{K_1 + K_2}{k_1 + k_2}$

We are given that $H_1, K_1, H_2, K_2, h_1, k_1, h_2, k_2$ are non-negative integers, and that

$$h_1 + k_1 \leqslant 10, \tag{1}$$

$$h_2 + k_2 \leqslant 10, \tag{2}$$

$$H_1/h_1 < K_1/k_1, \tag{3}$$

$$H_2/h_2 < K_2/k_2, \tag{4}$$

$$\frac{H_1 + H_2}{h_1 + h_2} > \frac{K_1 + K_2}{k_1 + k_2}\,. \tag{5}$$

Some people have an intuitive feeling that the three equations (3), (4), (5) can not be simultaneously true. The main reason for setting this problem is to demonstrate that such 'intuition' can be misleading.

Particular solution I

In the particular problem set (and using the notation of the General comment) we have that h_1, h_2, k_1, k_2 are non-negative integers; that (since 252/84 = 3)

$$h_1 + k_1 \leqslant 10, \tag{1}$$

$$h_2 + k_2 \leqslant 10, \tag{2}$$

$$h_1 > 3k_1, \tag{3}$$

$$3h_2 > k_2, \tag{4}$$

$$k_1 + k_2 > h_1 + h_2; \tag{5}$$

and also ('Ken took the first wicket in the first innings')

$$k_1 \geqslant 1. \tag{6}$$

By (4) $\quad 3h_2 - k_2 \geqslant 1$

and by (5) $\quad k_2 - h_2 \geqslant h_1 - k_1 + 1,$

so (adding the first of these to twice the second)

$$h_2 + k_2 \geqslant 2(h_1 - k_1) + 3.$$

So by (2)

$$h_1 - k_1 \leqslant 3.$$

But by (3), (6)

$$h_1 - k_1 \geqslant 3.$$

Hence

$$h_1 - k_1 = 3; \tag{7}$$

and, again by (3), (6), it follows that

$$\left.\begin{aligned} h_1 &= 4, \\ k_1 &= 1. \end{aligned}\right\} \tag{8}$$

Now by (2), (5), (7) we have that $h_2 \leqslant 3$; and by (4), (5), (7) we have that $h_2 \geqslant 3$. Hence

$$h_2 = 3. \tag{9}$$

Lastly, by (2), (9) we have that $k_2 \leqslant 7$; and by (5), (7), (9) we have that $k_2 \geqslant 7$. Hence

$$k_2 = 7. \tag{10}$$

Consequently the complete solution is as given in Table 4.2.

Table 4.2

	Bowler	Harry	Ken
1st innings	Runs against	252	84
	Wickets	4	1
	Average	63	84
2nd innings	Runs against	84	252
	Wickets	3	7
	Average	28	36
Whole match	Runs against	336	336
	Wickets	7	8
	Average	48	42

Particular solution II

In the particular problem set we have that h_1, h_2, k_1, k_2, K_1 are non-negative integers; that

$$h_1 + k_1 \leqslant 10, \tag{1}$$

$$h_2 + k_2 \leqslant 10, \tag{2}$$

$$h_1 > \xi k_1, \tag{3}$$

$$\xi h_2 > k_2, \tag{4}$$

$$k_1 + k_2 > h_1 + h_2, \tag{5}$$

where

$$\xi = 70/K_1 > 1; \tag{6}$$

and also ('In the first innings Ken took the first two wickets')

$$k_1 \geqslant 2. \tag{7}$$

Eliminating ξ between (3), (4) we have

$$h_1 h_2 > k_1 k_2, \tag{8}$$

and so by (1), (2)

$$(10 - k_1)(10 - k_2) > k_1 k_2,$$

from which

$$k_1 + k_2 \leqslant 9. \tag{9}$$

We also have from (3), (4), (5) that

$$k_1 + k_2 > \xi k_1 + k_2/\xi + 1,$$

from which

$$k_2 > \xi k_1 + \frac{\xi}{\xi - 1} \, . \qquad (10)$$

It follows from (7), (9), (10) that

$$\left.\begin{aligned} k_1 &= 2, \\ k_2 &= 7. \end{aligned}\right\} \qquad (11)$$

By (5), (8), (11) we then have that (h_1,h_2) are (3,5), (4,4), or (5,3). But by (2), (11) $h_2 \leqslant 3$.

Hence

$$\left.\begin{aligned} h_1 &= 5, \\ h_2 &= 3. \end{aligned}\right\} \qquad (12)$$

Inserting the results (11), (12) in (3), (4) it follows (using (6)) that

$$\frac{7}{3} < \frac{70}{K_1} = \xi < \frac{5}{2} \, .$$

Consequently

$$28 < K_1 < 30;$$

and hence

$$K_1 = 29.$$

The complete solution is given in Table 4.3.

Composer's problem

We want to set a problem that will demonstrate that (3), (4), (5) (in the notation of the General comment) can be simultaneously true. Our first difficulty is the embarrassingly large number of variables at our disposal. For the sake both of ease of solution and of ease of composition we impose the restraint $H_1 = K_2$, $H_2 = K_1$.

Table 4.3

	Bowler	Harry	Ken
1st innings	Runs against	70	29
	Wickets	5	2
	Average	14	$14\frac{1}{2}$
2nd innings	Runs against	29	70
	Wickets	3	7
	Average	$9\frac{2}{3}$	10
Whole match	Runs against	99	99
	Wickets	8	9
	Average	$12\frac{3}{8}$	11

Even so, there are still 'too many' solutions; so we impose the further restraint that k_1 is not to be zero.

Our investigation is then that of the 'Particular solution II' as far as equation (10) — with the exception of equation (7), which is replaced by $k_1 \geqslant 1$.

It is clear from (10) that there are no solutions with $k_1 \geqslant 3$ (since that would imply $h_2 + k_2 > 10$, contrary to (2)).

With $k_1 = 2$ we have that there is just one solution:

$$(h_1, h_2, k_1, k_2) = (5, 3, 2, 7);$$

and that for it to exist it is also necessary that

$$\frac{7}{3} < \xi < \frac{5}{2}.$$

This leaves for examination the case $k_1 = 1$. A simple and not too lengthy inspection shows that the possible solutions with $k_1 \geqslant 1$ (with their associated restraints on ξ) are as shown in Diagram 4.1.

We now want to select a set of values that minimizes the amount of data that needs to be provided to the solver, but still yields a unique solution. The values of ξ for which there is just one solution are

$$3/2 < \xi < 5/3, \quad 2 < \xi < 7/3, \quad 3, \quad 4 \leqslant \xi < 5.$$

Of these $\xi = 3$ seemed to me to be the most interesting: so I chose it. (The values for 'runs scored against' were then chosen so that the six bowling averages would all be integers.)

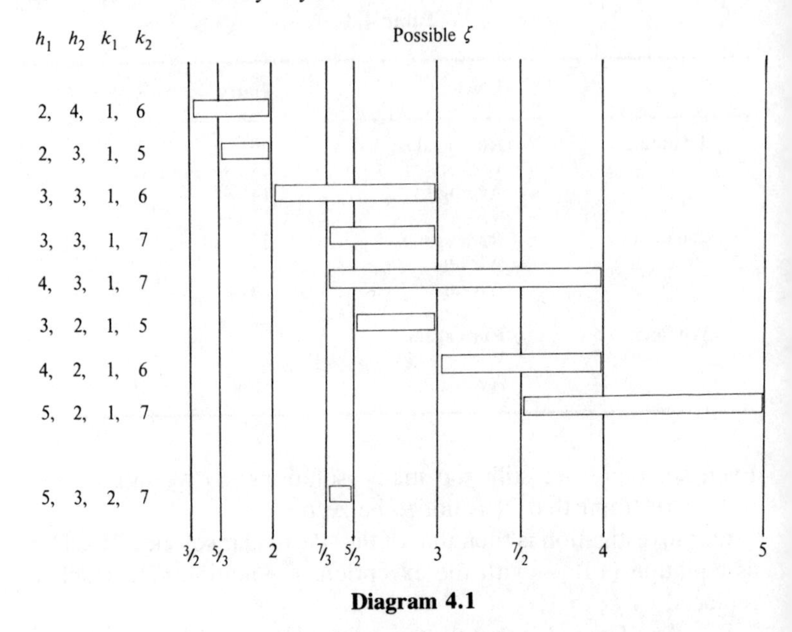

Diagram 4.1

The fact that there is only one solution for which $k_1 = 2$, and the rather short range of ξ for which this solution exists, suggested 'Bowling averages II' — which had not been in my mind when I first wrote 'Bowling averages I'.

5

CENTRE-POINT†

Problem

The villages of Ayling, Beeling, and Ceiling lie (as you will recall) in a plain. The County Council has just noticed that they are completely unconnected by road, and has decided to rectify that situation — with as short a total road length as possible.

Tom, Dick, and Harry have a map of the district with the three villages marked on it as points (A, B, C); they have been told to draw on it the road system that meets the Council's requirement.

Tom's proposal on how to do so involved two parallel sheets of glass held together (or, rather, held apart) by three pins whose positions correspond to the positions of the villages on the map. 'Plunge the contraption into soapy water; pull it out and let it settle; with luck there'll be a soap film between the sheets that will give us just what we're after.'

'Very pretty — theoretically,' said Harry; 'a ten-second experiment that needs hours of preparation.' Tom went off in a huff.

Dick's proposal involved three equal weights, three lengths of light string, a thin board, some glue, and an awl. 'Stick the map to the board; bore holes in it at the sites of the three villages. Tie a weight to each string; thread the other ends of the strings, one through each hole, from the back of the map; tie the three ends together in a knot in front of the map. Hold the board so that the weights hang clear; the strings on the front of the map will finish up along the roads we want.'

'Better,' said Harry; 'but while you're getting your kit together just leave me the map, will you, and I'll have the result marked in by the time you've found your string, never mind three equal weights and an awl.' Dick went off in a huff.

Harry had a pair of compasses, a straight-edge, a pen, and a pencil. He made a number of pencil marks (either straight lines or

† The title is a clue — but don't put too much faith in it.

arcs of circles) on the map. He then drew in ink the lines of the roads meeting the Council's requirement.

Naturally he made as few pencil marks as possible.

How many?

General solution

1.

F, G, H are given points.† We wish to connect them by a network of line-segments such that the network has minimum total length.

Without loss of generality we can assume that

$$GH \geqslant HF, FG.$$

Clearly the minimum network consists either of three line-segments (PF, PG, PH, where P is some point in or on the boundary of FGH); or of two line-segments (HF, FG). This second possibility can, however, be thought of as the particular case $P \equiv F$ of the first possibility: if we allow — as we shall — that PF may be of zero length then the first possibility is the only one that need be considered.

2.

Let H′ be the point (on the opposite side of FG to H) such that FGH′ is equilateral: see Diagram 5.1.

Let Q be *any* point in or on the boundary of FGH, and let Q′ be the point (on the opposite side of GQ to H) such that QGQ′ is equilateral.

The triangles QGF, Q′GH′ are congruent (two sides and included angle), and so $H'Q' = FQ$. We already have, by construction, that $Q'Q = GQ$. Hence

$$FQ + GQ + HQ = H'Q' + Q'Q + QH. \qquad (1)$$

We now consider separately the alternative situations

$$\widehat{GFH} \gtreqless 120°.$$

† Why 'F, G, H' rather than 'A, B, C'? Simply, to avoid overidentification with Ayling, Beeling, and Ceiling. F, G, H are going to turn out to be A, B, C in some order — but we do not yet know which order.

3. $\widehat{GFH} \geqslant 120°$

When $\widehat{GFH} \geqslant 120°$ the line HH′ does not cut FG internally (see Diagram 5.1(a)). Hence

$$\begin{aligned} H'Q' + Q'Q + QH &\geqslant H'Q + QH \\ &\geqslant H'F + FH, \end{aligned}$$

with equality only when $Q \equiv F$.

It follows by (1) (recalling that $GF = H'F$) that

$$FQ + GQ + HQ \geqslant GF + FH, \qquad (2)$$

with equality only when $Q \equiv F$.

Since (2) holds for *all* points Q in or on the boundary of FGH, we have that the minimum network consists of the two line-segments GF, FH.

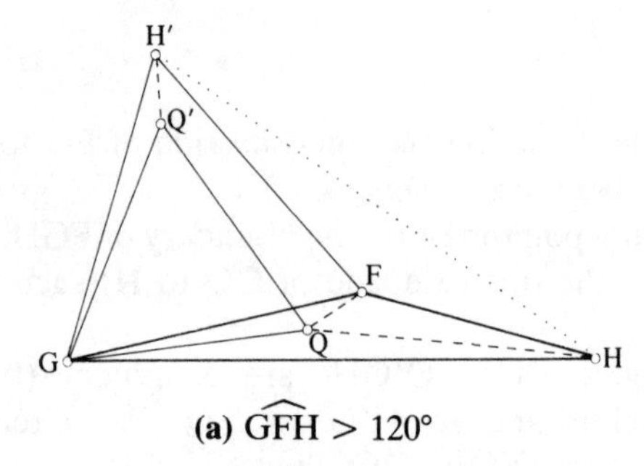

(a) $\widehat{GFH} > 120°$

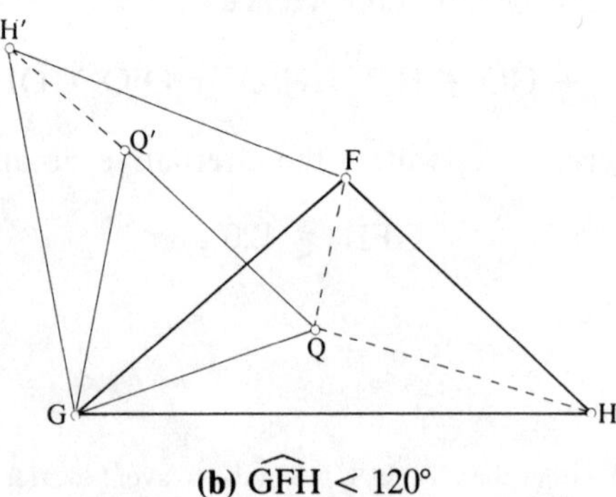

(b) $\widehat{GFH} < 120°$

Diagram 5.1

4. $\widehat{GFH} < 120°$

Let G′ be the point (on the opposite side of FH to G) such that FHG′ is equilateral. Let P be the point of intersection of GG′, HH′. (Note that when $\widehat{GFH} < 120°$ we have that P is inside FGH.) (See Diagram 5.2.)

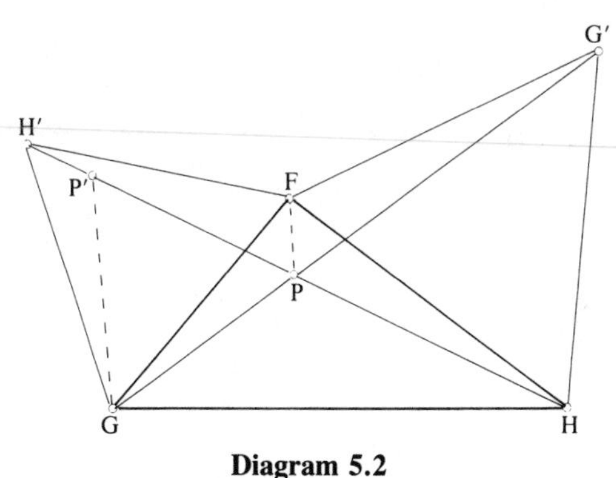

Diagram 5.2

The triangles GFG′, H′FH are congruent (two sides and included angle) and so $\widehat{FGG'} = \widehat{FH'H}$; that is, $\widehat{FGP} = \widehat{FH'P}$. Consequently FPGH′ is a cyclic quadrilateral, and so

$$\widehat{H'PG} = \widehat{H'FG} = 60°.$$

There is, then, a point P′ on HH′, between P and H′, such that PGP′ is equilateral.

The triangles PGF, P′GH′ are congruent (two sides and included angle), and so H′P′ = FP. We already have, by construction, that P′P = GP. Hence

$$\begin{aligned} FP + GP + HP &= H'P' + P'P + PH \\ &= H'H. \end{aligned} \tag{3}$$

From (1) we know that for all points Q in or on the boundary of FGH

$$FQ + GQ + HQ \geqslant H'H,$$

and from (3) we have that equality is attained when Q ≡ P.

Hence the minimum network consists of the three line-segments FP, GP, HP, where P is the point of intersection of GG′ and HH′.

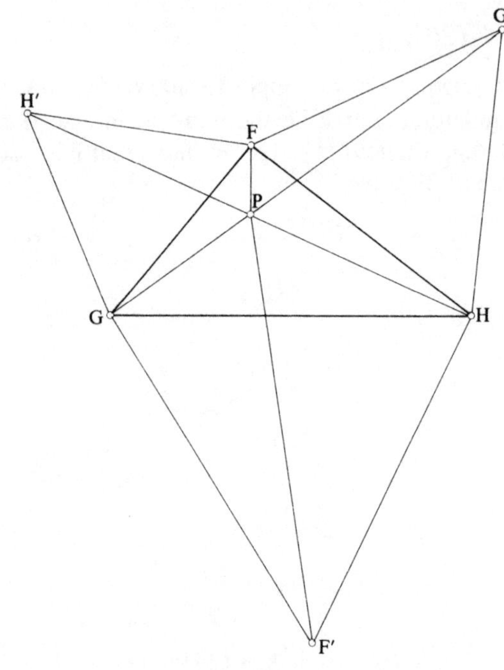

Diagram 5.3

5. $G\widehat{F}H < 120°$ (*continued*)

We could stop there: we have, after all, uniquely identified the minimum network: but it is probably desirable to go one stage further.

Let F′ be the point (on the opposite side of GH to F) such that GHF′ is equilateral. (See† Diagram 5.3.) Join F′P.

We already have that FPGH′ is a cyclic quadrilateral, and so $H'\widehat{P}F = H'\widehat{G}F = 60°$ and $H'\widehat{P}G = H'\widehat{F}G = 60°$. We similarly have that $G'\widehat{P}F = 60°$ and $G'\widehat{P}H = 60°$. Hence $G\widehat{P}H = 120°$; and so, since $G\widehat{F'}H = 60°$, we have that GPHF′ is a cyclic quadrilateral. Hence

$$G\widehat{P}F' = G\widehat{H}F' = 60°.$$

It follows that the six angles at P are each 60°, and, in particular,

† We are going to prove that F′, P, F are collinear. The diagram has been slightly distorted to help us avoid the error of assuming that they are before we've actually proved it.

that F, P, F′ are collinear. That is, the three lines FF′, GG′, HH′ have† a common point.

6.

We can now restate the result — without the assumption that we have previously made (that GH $\geqslant$ HF, FG).

I. If any angle of FGH equals or exceeds 120°, then the minimum network consists of the two sides of FGH that include that angle.

II. If all angles of FGH are less than 120°, then the minimum network consists of the three line-segments FP, GP, HP, where P is the common point of FF′, GG′, HH′.

Particular solution

Harry drew two arcs, with radius AB and centres A and B, meeting in C′ (on the opposite side of AB to C); and two arcs, with radius AC and centres A and C, meeting in B′ (on the opposite side of AC to B).

He then drew a straight line through B, B′, and a straight line through C, C′; these two lines meeting in P.

That completed his construction (Diagram 5.4): he had made *six* pencil marks.

His final step (Diagram 5.5) is to draw‡ the roads:

(i) If P is inside ABC: three roads — AP, BP, CP.

(ii) If P is outside ABC: two roads — two of BC, CA, AB, leaving out the longest of them.

Composer's problem

The difficulty here lies not so much in the setting of the problem as in the presentation of its solution.

† This is also true when $\widehat{GFH} > 120°$; but in that case P is outside FGH, and the proof needs some slight recasting. I do not include the recast proof, since we do not need it for this problem.

‡ Strictly speaking, we need to show that Harry's pencil construction is sufficient to tell him *which* set of roads to draw.

(a) If P is on B′B extended (rather than between B and B′): the roads are AB, BC.
(b) If P is on C′C extended (rather than between C and C′): the roads are AC, CB.
(c) If A is an interior point of the region bounded by PB (extended beyond B) and PC (extended beyond C): the roads are BA, AC.
(d) Otherwise: the roads are PA, PB, PC.

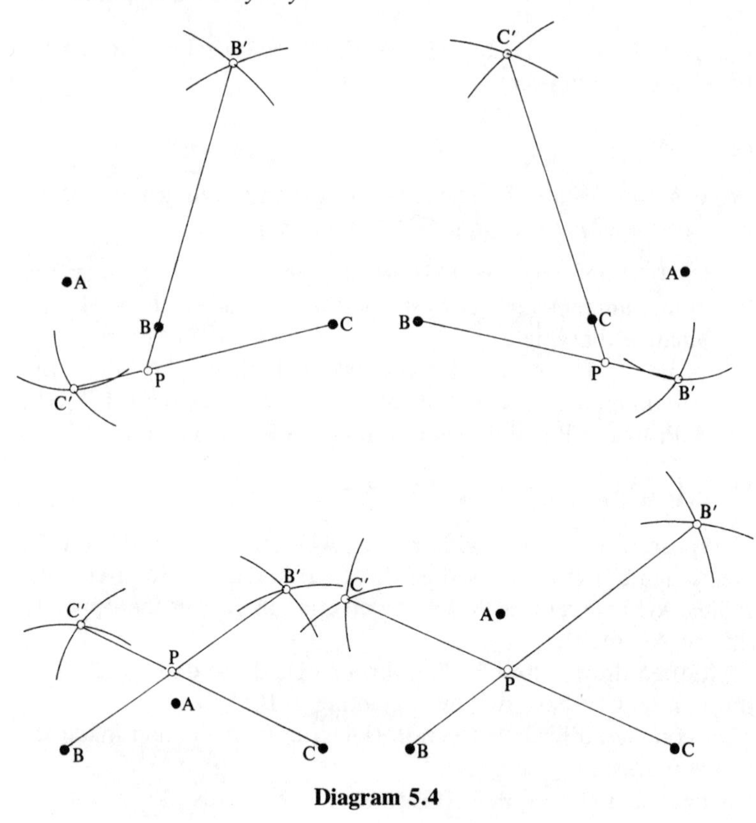

Diagram 5.4

There is an interesting geometrical theorem:

> If one erects equilateral triangles A′BC, AB′C, ABC′ on the sides of a given triangle ABC, and external to it, then the lines AA′, BB′, CC′ have a common point, P; and, further, if P is internal to ABC then $\widehat{BPC} = \widehat{CPA} = \widehat{APB} = 120°$.

In tackling the minimum network problem it is tempting to start by proving that theorem. But, in one sense, the theorem goes too far (we do not need the fact that the three lines AA′, BB′, CC′ have a common point: we just need to identify the common point of two of them); and, in another sense, it does not go far enough (it does not help us in the case in which one of the angles of ABC exceeds 120°).

So I felt that the temptation should be resisted; that the

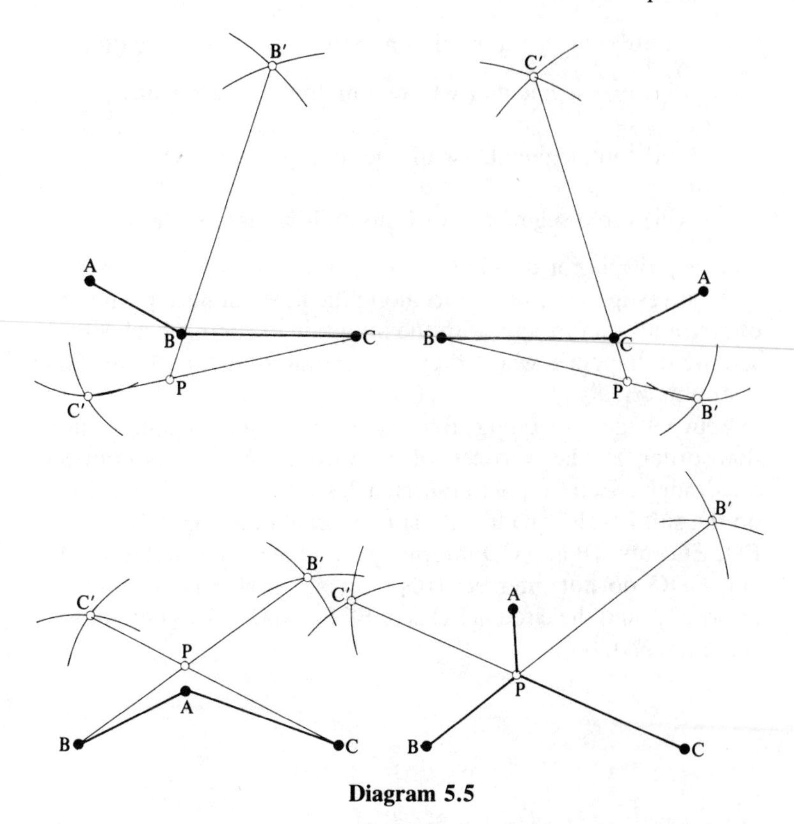

Diagram 5.5

minimum network problem should be tackled from scratch; and that the theorem described above should be allowed to come out almost as a corollary to the network result (rather than the other way round).

Extension

What happens it, instead of three villages, there are four?

With three villages the minimum network consists of either

(i) three segments (with one multiple intersection),

or

(ii) two segments (with no multiple intersection);

and the criterion for deciding between these alternatives is simple: 'does the triangle ABC have an angle exceeding 120°?'

With four villages the minimum network consists of either

(i) five segments (with two multiple intersections),

or

(ii) four segments (with one multiple intersection),

or

(iii) three segments (with no multiple intersection);

and the criteria for deciding between these are not so simple.

An investigation might start along the lines (imposing a number of restraints on the way, with the intention of coming back later to see what happens when they are removed: let us label these restraints $\mathscr{R}_1, \mathscr{R}_2, \mathscr{R}_3, \ldots$):

Four villages — Ayling, Beeling, Ceiling, and Dealing — lie in that order at the vertices of a convex (this is restraint $\mathscr{R}_1$) quadrangle. Erect equilateral triangles ABE, BCF, CDG, DAH on the sides of the quadrangle and external to it; join EG and join FH. EG cuts AB and CD internally, and the circumcircles of ABE and CDG do not intersect (restraint $\mathscr{R}_2$). FH cuts BC and DA internally, and the circumcircles of BCF and DAH do not intersect (restraint $\mathscr{R}_3$).

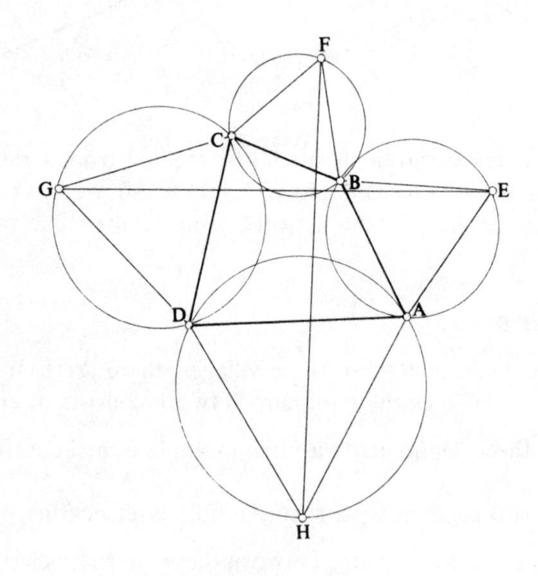

Diagram 5.6

(We seem to have imposed some fairly considerable restraints, but the kind of quadrangle that one tends to sketch usually meets them: see Diagram 5.6).

It is then fairly easy to show that the minimum network has total length EG or FH — whichever is the lesser.

Suppose EG is the lesser. Let EG cut the circumcircle of ABE in I, and the circumcircle of CDG in J. The five segments of the minimum network are AI, BI, IJ, CJ, DJ. (This of course requires proof; but the proof is straightforward — using the results for the minimum network in the case of a triangle.) See Diagram 5.7.

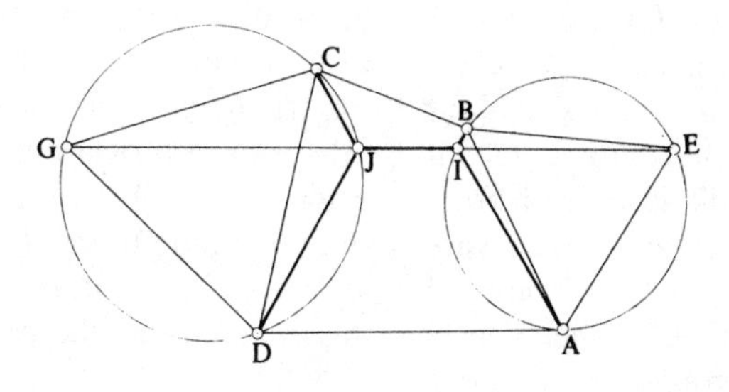

Diagram 5.7

If we think of a quadrangle that meets the three restraints $\mathscr{R}_1$, $\mathscr{R}_2, \mathscr{R}_3$ as being 'well-behaved', then we have that the minimum network for a well-behaved quadrangle has five segments (with two multiple intersections), and we have a simple way in which to construct the network. It does *not* follow that every quadrangle that has a minimum network with five segments is well behaved: removal of the restraint $\mathscr{R}_3$, for example, can still lead to quadrangles whose minimum networks are of five segments — but can also lead to ones whose minimum networks are of four — or three — segments.

Just what are the necessary and sufficient restraints that we need to impose on the quadrangle to ensure that its associated minimum network is of five segments? And of four segments? And of three segments?

And, having done that, how about five villages? six villages? seven villages? ...

6

COUNTING SHEEP

Problem

Beeling is rather a quiet village — except on the day of the annual sheep market. On that day lines of hurdles are erected from the flag-pole to each of the corners of the market-place, and along three of its four sides, making three triangular pens, of different sizes. Each hurdle is a yard long, there are no overlaps or gaps, and no hurdle is bent or broken. Each pen is filled with sheep — one sheep to each square yard. (The remaining triangular area of the market-place is occupied by dealers and spectators.)

The number of sheep in each pen is equal to the number of hurdles surrounding that pen.

What is the area of the Beeling market-place?

Solution

Consider a triangular pen whose sides are formed by a, b, c hurdles; then the number of sheep in the pen is

$$a + b + c. \tag{1}$$

Since each hurdle is a yard long, and there is one sheep to each square yard, we also have that the number of sheep in the pen is

$$\tfrac{1}{4}\sqrt{\{(a+b+c)(-a+b+c)(a-b+c)(a+b-c)\}}. \tag{2}$$

Equating (1) and (2):

$$16(a+b+c) = (-a+b+c)(a-b+c)(a+b-c). \tag{3}$$

Now a, b, c are integers; and each of the factors on the right of (3) is positive. If one of these factors were odd then the other two would also be odd; so (because of the factor 16 on the left of (3)) all of them are even.

It follows that there are positive integers p, q, r such that

$$-a + b + c = 2p,$$

$$a - b + c = 2q,$$

$$a + b - c = 2r;$$

from which

$$\left.\begin{aligned} a &= q + r, \\ b &= r + p, \\ c &= p + q; \end{aligned}\right\} \tag{4}$$

and by (3)

$$4(p + q + r) = pqr. \tag{5}$$

By symmetry we can assume without loss of generality that

$$p \leqslant q \leqslant r. \tag{6}$$

If $p = 1$ we have by (5) that $(q - 4)(r - 4) = 20$, and so by (6) $(q,r) = (5,24), (6,14), (8,9)$.

If $p = 2$ we have by (5) that $(q - 3)(r - 2) = 8$, and so by (6) $(q,r) = (3,10), (4,6)$.

If $p = 3$ we have by (5) that $(3q - 4)(3r - 4) = 52$, which has no solution satisfying (6).

If $p \geqslant 4$ we have by (5) that $(q - 1)(r - 1) \leqslant 5$, which has no solution satisfying (6).

Inserting these results in (4) it follows that there are just five solutions of (3):

$$\left.\begin{array}{cccccc} a & 29 & 20 & 17 & 13 & 10 \\ b & 25 & 15 & 10 & 12 & 8 \\ c & 6 & 7 & 9 & 5 & 6 \end{array}\right\} \qquad (7)$$

The three pens actually used must be of different sizes, and one of them must have two sides each of which is common with another pen used. The only possibility (by inspection of (7)) is that this is the pen (10,8,6), having side 10 in common with (17,10,9) and side 6 in common with (29,25,6).

At first sight there are four possible cases to be considered — four possible shapes of the market-place — depending on 'which way round' the (17,10,9) and (29,25,6) pens fit to the (10,8,6) pen. These four cases are sketched in the Diagrams 6.1–6.4. In each of the four diagrams F represents the flag-pole and A,B,C,D represent the corners in order round the market-place.

In Cases I and II the areas of $\triangle$BCF, $\triangle$CDF are 24, 60 square yards, which are in the proportion 2 : 5. Since BF, DF are in the same proportion it follows that BFD is a straight line. Since the area of $\triangle$ABF is 36 square yards it follows that the area of $\triangle$DAF is $\frac{5}{2} \times 36 = 90$ square yards.

In cases I and III the areas of $\triangle$BCF, $\triangle$ABF are 24, 36 square yards, which are in the proportion 2 : 3. Since CF, AF are in the same proportion it follows that AFC is a straight line. Since the area of $\triangle$CDF is 60 square yards it follows that the area of $\triangle$DAF is $\frac{3}{2} \times 60 = 90$ square yards.

Case IV is unacceptable as a solution because it involves the area occupied by the pens being included in the area occupied by the dealers and spectators.

Consequently there are just three possible configurations, in each of which the total area of the market place is

$$36 + 24 + 60 + 90 = 210 \text{ square yards.}$$

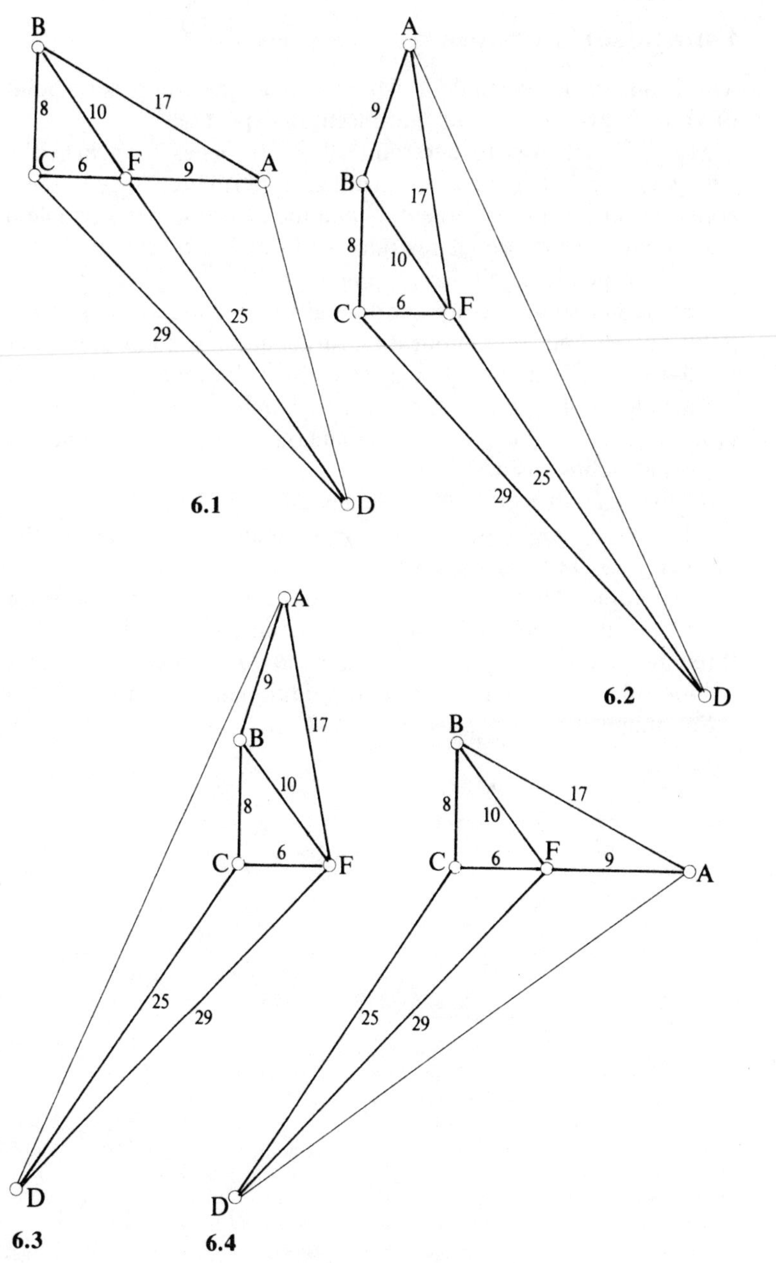

Diagrams 6.1–6.4

Composer's problem

This is one of those problems where — from the composer's point of view — everything falls into place: unexpectedly.

My first idea was to determine those triangles for which the (integer) area 'equalled' the (integer) perimeter — in some coherent unit of measurement — and then to compose a problem with whatever additional restraint was needed to cut down the number of possible solutions to one.

The actual solutions are the five given in equation (7) of the Solution — and the duplication of 6 and 10 (and no other numbers) in (7) naturally suggested the 'additional restraint' of being able to fit together three of the triangles: the three triangles (10,8,6), (17,10,9), (29,25,6) — and no other possible combination — would be involved.

At this stage the problem looked like being 'What are the areas of the three sheep pens?', or 'How many sheep are at the market?', or some such question.

Then came the realization that, of the four possible ways of fitting the pens together, one was 'inadmissible', and the other three all gave the same total area for the quadrilateral ABCD: a unique solution with a minimum of additional restraint.

Serendipity!

7

TRANSPORT

Problem I

Tom, Dick, and Harry wanted to prolong their visit to the Ayling Arms as much as possible, but they were due to meet their boss at 3.00 in his office 9 miles away, and knew that they mustn't be late for the appointment.

Tom had a bicycle and Dick had a scooter — but Harry had no transport at all. So they decided to share.

Each of them could travel on foot at 5 m.p.h.; each of them could scoot† at 9 m.p.h.; and each of them could cycle at 15 m.p.h. Neither the scooter nor the bicycle could carry more than one person; but each could safely be left unattended by the roadside (they're an honest lot in those parts).

The three of them left the Ayling Arms together, at the last possible moment; the three of them arrived together just in time for their 3.00 appointment.

At what time did they leave the Ayling Arms?

† Added in proof. A well-meaning friend suggests to me that 9 m.p.h. is unrealistically slow for a scooter. He is in error. When I say 'scooter' I mean 'scooter'; I do not mean 'motor-scooter'.

Problem II†

You will recall that Tom has a bicycle that travels at 15 m.p.h., Dick has a scooter that travels at 9 m.p.h., and Harry has no transport at all.

A week after their last joint expedition (Problem I), the three of them were joined at the Ayling Arms — 9 miles from the Chief Surveyor's office — by Ken (whose motorcycle travels at 30 m.p.h.), John (whose tricycle travels at 12 m.p.h.), and Brian and Ian (with no transport at all).

The Chief Surveyor has requested that all seven of them appear in his office at 3.00.

They decided to share the available transport (four vehicles, each capable of carrying just one person at its prescribed speed, and each of which can be left safely by the roadside when not in use, to be picked up later by another member of the group).

The seven of them left the Ayling Arms together, at the last possible moment; the seven of them arrive together just in time for their 3.00 appointment.

At what time did they leave the Ayling Arms?

† If you have already worked out the *general* approach to this kind of problem, then this example should present you with little difficulty — but if your solution to Problem I was a 'trial and error particular', then I suggest that you give this one a miss!

Problem III

After last week's arguments (about who should leave what where, with seven people and four vehicles involved) our three regulars — Tom, Dick, and Harry — decided to give the Ayling Arms a miss, and met at the Beeling Bistro (even though this was 10 miles from the Chief Surveyor's office).

Dick was feeling grim about his scooter; it had been damaged at the end of its journey the previous week, and now its speed was only 6 m.p.h.: he was fully prepared to throw it away, and couldn't care less what happened to it.

(Tom's bicycle still travels at 15 m.p.h.; Harry still has no transport at all; the three of them still each travel on foot at 5 m.p.h.)

As before, the three of them share the available transport; the three of them leave the Beeling Bistro together at the last possible moment; the three of them arrive together just in time for their 3.00 appointment with the Chief Surveyor.

At what time did they leave the Beeling Bistro?

General solution

A group of travellers want to get to a destination remote from their starting place. A number of vehicles are available to them. They cooperate in the use of the vehicles. What is the least time that must elapse before the last of them arrive at the destination?

In this sort of problem there are always more travellers than vehicles; each vehicle can carry just one person, and — when manned — moves at its own constant speed, irrespective of who is driving it; any unmanned vehicle remains stationary; and all travellers, when travelling on foot, travel at the same pace.

Usually the initial information is the length of the journey (L); the number of travellers (M), and the speed at which they travel on foot (v_0); the number of vehicles (N), and their speeds when manned ($v_1, v_2, \ldots, v_N$). The problem is to find the least possible value (T) of the time taken by the last traveller to arrive at the destination.

Probably the simplest approach is to work as far as possible in times, rather than in speeds: define

$$\left.\begin{aligned} t_0 &= L/v_0, \\ t_1 &= L/v_1, \\ t_2 &= L/v_2, \\ &\cdots\cdots \\ t_N &= L/v_N. \end{aligned}\right\} \tag{1}$$

(t_0 is the time it would take a traveller to cover the whole distance on foot; t_r ($r = 1, 2, \ldots, N$) is the time it would take to cover the whole distance using just the rth vehicle.)

If there were no vehicles available at all, the total of the times taken by the M travellers would be Mt_0. Making available a vehicle with speed v_r can — at best — reduce this total by $(t_0 - t_r)$. The actual availability of the N vehicles reduces the total by — at best —

$$(t_0 - t_1) + (t_0 - t_2) + \ldots + (t_0 - t_N);$$

that is, the total of the times of the M travellers is — at least —

$$(M - N)t_0 + t_1 + t_2 + \ldots + t_N.$$

Define

$$T^* = \{(M - N)t_0 + t_1 + t_2 + \ldots + t_N\}/M. \tag{2}$$

Then we have

$$T \geqslant T^*, \tag{3}$$

where T is the time taken by the last traveller to arrive at the destination.

We also have, incidentally, that if T is as little as T^* then every one of the travellers takes precisely T^* to get to the destination (because T^* is the average of the times they take, and T is the time taken by the last arrival); and that

$$t_1, t_2, \ldots, t_N \leqslant T^* \tag{4}$$

(because it is necessary to take full advantage of every vehicle for the whole length of the journey — and if any vehicle takes more than T^* to complete the journey, then its last user must also take more than T^*).

It is convenient at this stage to distinguish three cases: one of them 'Fast' {in which the slowest of the vehicles is still sufficiently fast for (4) to be satisfied}, and two of them 'Slow' {in which (4) is not satisfied}.

'Fast'

We know by (3) that T can not be less than T^*, but is there always a pattern of cooperative vehicle usage by the travellers that allows T to be as little as T^*? The answer to that question is interestingly straightforward: 'Yes — if and only if (4) is satisfied.'

A proof of this result is, in a sense, equally straighforward — it consists simply of a detailed description of a pattern of vehicle usage that meets the requirement that no traveller takes longer than T^* to get to the destination. Unfortunately descriptions of such patterns — in the general case — seem to be somewhat complicated. The best (or, rather, the least bad) that I have found I have accordingly dismissed to an Appendix, rather than include it here. (An idea of the pattern can be obtained from the two particular cases illustrated in the solutions to Problems I and II.)

'Slow'

From now on we look solely at the cases in which (4) is *not* satisfied: one at least of the vehicles would take more than T^* to complete the journey, even if driven continuously all the way.

There are two possible problems, depending on the answer to the question: 'Is it allowable to abandon vehicles on the way?'

'Slow A'

If vehicles may not be abandoned on the way, then the value of T can again be simply stated: it is the time it takes the slowest vehicle to complete the journey, driven continuously all the way: i.e.

$$T = \max(t_1, t_2, \ldots, t_N). \tag{5}$$

The proof of (5) is straightforward. Relabel the vehicles, if necessary, so that

$$t_1 \leqslant t_2 \leqslant \ldots \leqslant t_{N-1} \leqslant t_N.$$

Clearly T cannot be less than t_N. Allocate one traveller to drive the slowest vehicle all the way; he takes time t_N.

We are left with $M - 1$ travellers with $N - 1$ vehicles between them: a 'reduced' transport problem. This reduced problem may be either a 'Fast' one or a 'Slow A' one.

If it is a 'Fast' problem, then we know that each of the $M - 1$ travellers can get to the destination in time

$$\{(M - N)t_0 + t_1 + t_2 + \ldots + t_{N-1}\}/(M - 1),$$

which is less than t_N.

If on the other hand the reduced problem is a 'Slow A' one, then we go through the process again: sending one traveller on the slowest of the $N - 1$ vehicles, taking time t_{N-1} (which is less than or equal to t_N), and then examining the remaining problem of $M - 2$ travellers with $N - 2$ vehicles.

We may have to go through this process several times, but eventually we will get to a 'Fast' problem — even if it is one in which there is only one vehicle.

'Slow B'

If vehicles can be abandoned on the way, then we can improve on (5). But by how much?

When there are three travellers and two vehicles, as in Problem III, then the solution is fairly easy. But when the number of travellers exceeds the number of vehicles by two or more, and we look at the general case, then the solution is (at the time of writing!) beyond me.

Particular solution I

In the notation of the General solution, we have

$$L = 9 \text{ miles},$$

$$\left.\begin{array}{l} v_0 = 5 \text{ m.p.h.} \\ v_1 = 15 \text{ m.p.h.} \\ v_2 = 9 \text{ m.p.h.} \end{array}\right\} \quad \text{and so} \quad \left\{\begin{array}{l} t_0 = 108 \text{ minutes}, \\ t_1 = 36 \text{ minutes}, \\ t_2 = 60 \text{ minutes}. \end{array}\right.$$

Hence

$$\begin{aligned} T^* &= \{(3-2)108 + 36 + 60\}/3 \text{ minutes} \\ &= 68 \text{ minutes}. \end{aligned}$$

Since 36 and 60 are both less than 68 it follows that the problem is a 'Fast' one, and consequently that

$$T = 68 \text{ minutes}.$$

Tom, Dick, and Harry arrived together at 3.00 at their destination, so the last possible moment at which they could have left the Ayling Arms is 1.52.

There are many ways in which they could have cooperated to achieve this; one of them is indicated in Diagram 7.1.

Particular solution II

In the notation of the General solution, we have

$$L = 9 \text{ miles},$$

$$\left.\begin{array}{l} v_0 = 5 \text{ m.p.h.} \\ v_1 = 30 \text{ m.p.h.} \\ v_2 = 15 \text{ m.p.h.} \\ v_3 = 12 \text{ m.p.h.} \\ v_4 = 9 \text{ m.p.h.} \end{array}\right\} \quad \text{and so} \quad \left\{\begin{array}{l} t_0 = 108 \text{ minutes}, \\ t_1 = 18 \text{ minutes}, \\ t_2 = 36 \text{ minutes}, \\ t_3 = 45 \text{ minutes}, \\ t_4 = 60 \text{ minutes}. \end{array}\right.$$

Hence

$$\begin{aligned} T^* &= \{(7-4)108 + 18 + 36 + 45 + 60\}/7 \text{ minutes} \\ &= 69 \text{ minutes}. \end{aligned}$$

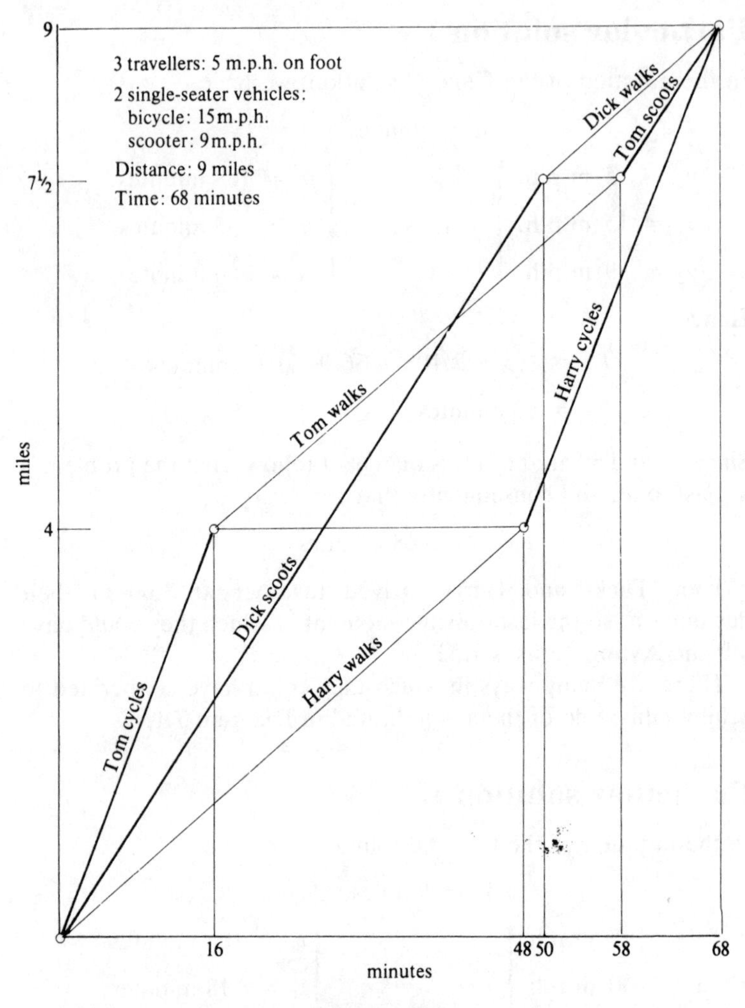

Diagram 7.1

Since 18, 36, 45, 60 are all less than 69 it follows that the problem is a 'Fast' one, and consequently that

$$T = 69 \text{ minutes.}$$

The seven travellers arrive together at 3.00 at their destination, so the last possible moment at which they could have left the Ayling Arms is 1.51.

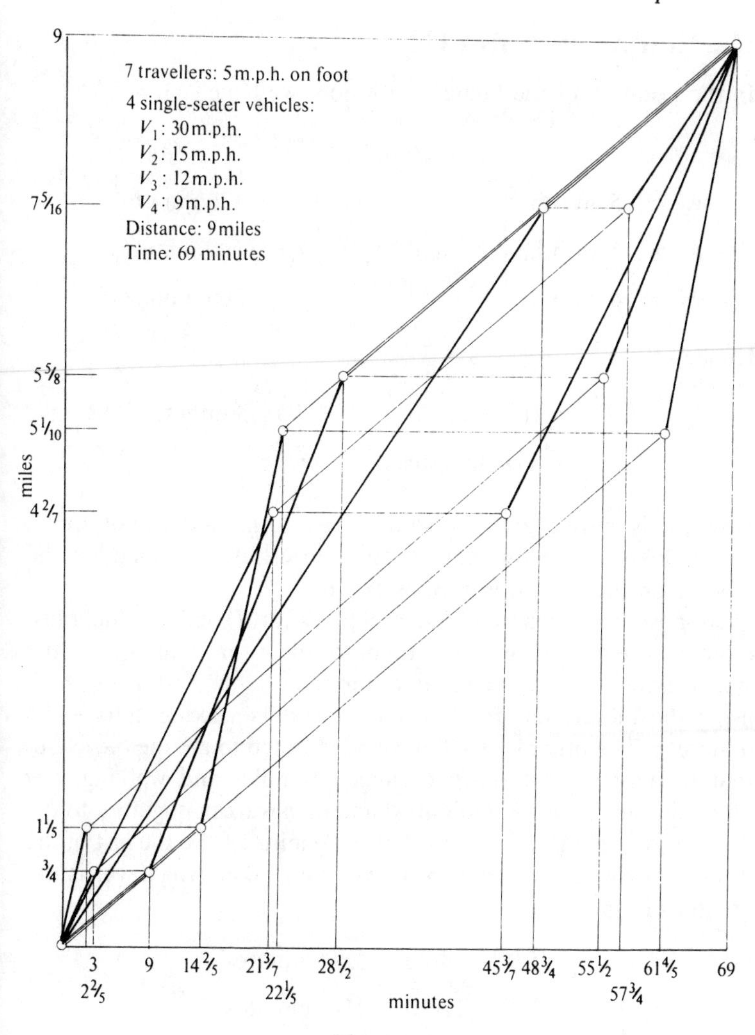

Diagram 7.2

One possible way in which they could have cooperated to achieve this is indicated in Diagram 7.2.

Particular solution III

In the notation of the General solution, we have

$$L = 10 \text{ miles},$$

$$\left.\begin{aligned} v_0 &= 5 \text{ m.p.h.} \\ v_1 &= 15 \text{ m.p.h.} \\ v_2 &= 6 \text{ m.p.h.} \end{aligned}\right\} \quad \text{and so} \quad \left\{\begin{aligned} t_0 &= 120 \text{ minutes}, \\ t_1 &= 40 \text{ minutes}, \\ t_2 &= 100 \text{ minutes}. \end{aligned}\right.$$

Hence

$$\begin{aligned} T^* &= \{(3-2)120 + 40 + 100\}/3 \text{ minutes} \\ &= 86\tfrac{2}{3} \text{ minutes}. \end{aligned}$$

Now 100 is more than $86\frac{2}{3}$, so the 'Fast' formula does not apply: the problem is a 'Slow' one — and, since they can abandon the scooter on the way, it is a 'Slow B' one.

Suppose they abandon the scooter after D miles. The most effective use of it is for one of them to ride it for D miles from the start (taking $10\,D$ minutes), and then to cycle the rest of the way, using the bicycle left for him by the others {taking $4(10 - D)$ minutes}. The other two will have cooperated to get the bicycle to that take-over point (each cycling $\frac{1}{2}D$ miles and walking $\frac{1}{2}D$ miles, so each taking $8D$ minutes) and then walking together to the destination, taking a further $12(10 - D)$ minutes. So the last of the three to arrive at the destination will have taken whichever is the greater of

$$\begin{aligned} 10D + 4(10 - D) &\quad \text{minutes}, \\ 8D + 12(10 - D) &\quad \text{minutes}. \end{aligned}$$

This is least when these two amounts are equal; that is, when $D = 8$.

With this value of D we have that the common time for the three travellers to get to their destination is 88 minutes.

Tom, Dick, and Harry arrived together at 3.00 at their destination, so the last possible moment at which they could have left the Beeling Bistro is 1.32.

This solution is indicated in Diagram 7.3.

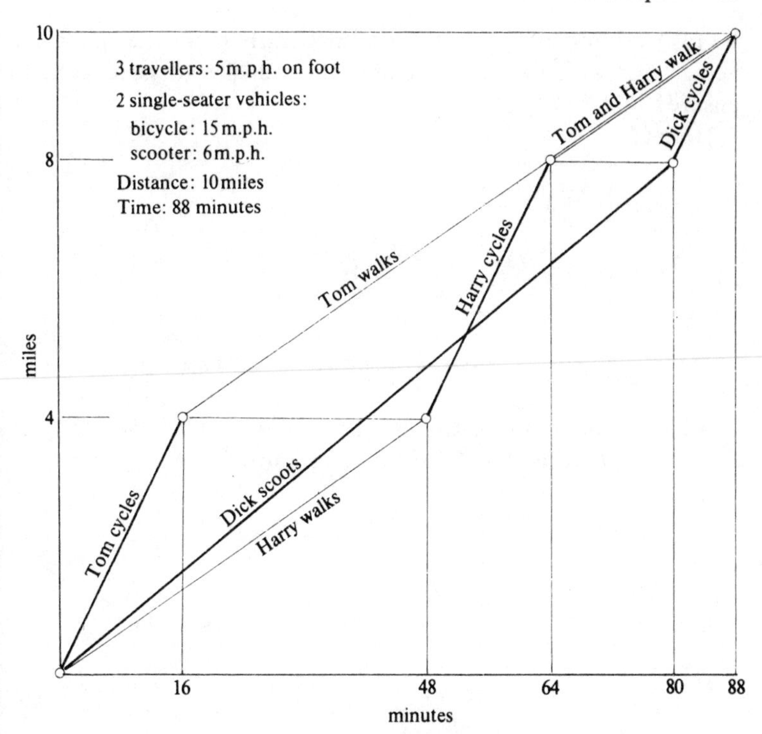

Diagram 7.3

Appendix

In the General solution it was asserted that, if

$$t_1, t_2, \ldots, t_N \leqslant T^*, \tag{4}$$

where

$$T^* = \{(M - N)t_0 + t_1 + t_2 + \ldots + t_N\}/M,$$

then

$$T = T^*,$$

where T is the time taken by the last traveller (and, incidentally, by all the other travellers) to arrive at the destination.

Here I describe a pattern of cooperative vehicle usage that validates that assertion.

Define

$$\Omega_s = \frac{T^* - t_s}{t_0 - T^*} \qquad (s = 1, 2, \ldots, N);$$

by (4) we have that $\Omega_s \geqslant 0$ for all s. Clearly there is at least one vehicle for which $\Omega_s > 0$; relabel the vehicles, if necessary, to ensure† that $\Omega_N > 0$.

Define‡

$$A_s = \begin{cases} [\Omega_1] & (s = 1) \\ \left[\sum_{u=1}^{s} \Omega_u\right] - \left[\sum_{u=1}^{s-1} \Omega_u\right] & (s = 2, 3, \ldots, N-1) \\ M - N - 1 - \left[\sum_{u=1}^{N-1} \Omega_u\right] & (s = N) \end{cases}$$

All the A_s are non-negative integers, and their sum is $M - N - 1$. Recall that L is the length of the total journey.

Define

$$J_s = \frac{L}{1 + \Omega_s} \qquad (s = 1, 2, \ldots, N)$$

$$K_s = \begin{cases} J_1 & (s = 1) \\ J_s\left\{1 + \left[\sum_{u=1}^{s-1} \Omega_u\right] - \sum_{u=1}^{s-1} \Omega_u\right\} & (s = 2, 3, \ldots, N). \end{cases}$$

We now label the travellers:

(i) Allocate $M - N - 1$ of them to N subgroups of sizes $A_1, A_2, \ldots, A_N$. (Note that any of these subgroups may be of size zero: indeed, if $M = N + 1$, then all the subgroups are of size zero.) Label the travellers in the sth subgroup P(s,1), P(s,2), …, P(s,A_s).

(ii) Label the remaining $N+1$ travellers Q(1), Q(2), … , Q($N+1$).

Let V_s be the vehicle with speed v_s.

We now tell the travellers what to do:

P(s,u): travel on foot a distance $K_s + (u - 1)J_s$, then travel on V_s a distance J_s, then travel on foot the rest of the way.

† If we did not take this precaution, the following definition could involve '$A_N = -1$'.

‡ $[x]$ denotes the integer part of x.

Q(1): travel on V_1 a distance K_1, then travel on foot the rest of the way.

Q(s) ($s = 2, 3, \ldots, N$): travel on V_s a distance K_s, then travel on foot a distance $A_{s-1}J_{s-1} + K_{s-1} - K_s$, then travel on V_{s-1} the rest of the way.

Q($N + 1$): travel on foot a distance $A_N J_N + K_N$, then travel on V_N the rest of the way.

This pattern of vehicle usage is more easily described from the point of view of the vehicles (rather than of the travellers).

V_s starts the journey with Q(s), who leaves it K_s from the start. V_s is then used by the A_s travellers P(s,u) successively, each of them taking it a distance J_s. V_s is then used by Q($s + 1$), who takes it to the final destination.

Verification that each traveller takes precisely T^* to complete the journey (and that none of the vehicles has a negative waiting time for its next user!) is a little tedious, but straightforward.

Extension

I spent more time than I care to admit to before I realized that T (the time taken by the last traveller to arrive at the destination) could not be less than T^* (the average of the times that the travellers would take if each of them travelled independently — N of them each using one vehicle all the way, and the other $M - N$ of them travelling on foot all the way).

Clearly

$$t_1, t_2, \ldots, t_N \leqslant T^* \tag{4}$$

is a necessary condition for $T = T^*$ — but that it is a sufficient one is not so obvious. My approach was to find an actual pattern of vehicle usage that (given (4)) provided $T = T^*$; but the pattern that I describe in the Appendix — while being the simplest that I have found — is still undesirably complicated.

My first 'Extension question is, therefore:

Is there a short simple proof that $T = T^*$ when (4) is satisfied?

(Shorter and simpler, that is, than the one given in the Appendix.)

My second, and major, 'Extension' question is:

What is the value of T when (4) is not satisfied — and vehicles can be abandoned?

I have found this question to be completely intractable (in the general case).

8

ALLEY LADDER

Problem

Dick and Harry are (part-time) window cleaners. On one occasion they were working together in Ayling Alley (which is a horizontal road with parallel vertical walls on either side). Dick set his ladder with its foot at the base of the south wall, leaning squarely† against the north wall; Harry set his ladder with its foot at the base of the north wall, leaning squarely against the south wall. The two ladders just touched each other.

Dick's ladder is 3 ft longer than Harry's ladder, and reaches 4 ft higher up the north wall than Harry's does up the south wall. The point at which the two ladders touch each other is 5 ft 10 in. above the ground.

How wide is Ayling Alley?

† By 'squarely' I mean that the vertical projection of the ladder is at right angles to the wall.

General solution

Let the width of the alley be β. One ladder, of length γ_1, has its foot on one side of the alley and reaches the opposite wall at height α_1. The other ladder, of length γ_2, reaches its opposite wall at height α_2.

We have

$$\alpha_1^2 + \beta^2 = \gamma_1^2, \tag{1}$$

$$\alpha_2^2 + \beta^2 = \gamma_2^2. \tag{2}$$

The two ladders touch each other at height δ above the ground (see Diagram 8.1).

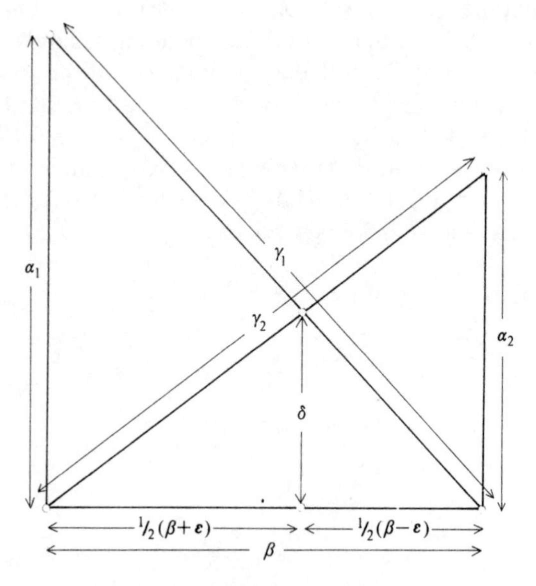

Diagram 8.1

By similar triangles we have

$$\frac{\beta}{\alpha_1} = \frac{\frac{1}{2}(\beta - \varepsilon)}{\delta},$$

$$\frac{\beta}{\alpha_2} = \frac{\frac{1}{2}(\beta + \varepsilon)}{\delta};$$

from which

$$\frac{1}{\alpha_1} + \frac{1}{\alpha_2} = \frac{1}{\delta}.$$

First Type of Problem

We have six unknowns (α_1, α_2, β, γ_1, γ_2, δ) connected by the three equations (1), (2), (3). Given the values of any three (independent) unknowns we can — theoretically — calculate the values of the other three. This calculation is easy (it does not even involve the solution of a quadratic) in every case except one: that one case is the situation in which we are given the values of γ_1, γ_2, δ and asked to obtain the value of any one of α_1, α_2, β. However we manipulate the equations, we are in this case eventually faced with having to solve a quartic equation.

Second Type of Problem

We may, however, not be given the values of three of the six unknowns directly, but, rather, three relationships between them. Most frequently these are sums or differences of α_1, α_2 and γ_1, γ_2.

For convenience we write

$$\left.\begin{aligned} \alpha_1 + \alpha_2 &= p, \quad \gamma_1 + \gamma_2 = r, \\ \alpha_1 - \alpha_2 &= q, \quad \gamma_1 - \gamma_2 = s. \end{aligned}\right\} \tag{4}$$

Substituting from (4) in (1), (2), (3) we have

$$(p + q)^2 + 4\beta^2 = (r + s)^2, \tag{5}$$

$$(p - q)^2 + 4\beta^2 = (r - s)^2, \tag{6}$$

$$(p + q)(p - q) = 4p\delta. \tag{7}$$

We again have six unknowns (p, q, r, s, β, δ), connected by the three equations (5), (6), (7). Given the values of any three (independent) unknowns we can calculate the values of the other three. The calculation is easy (it does not even involve the solution of a quadratic) except in the cases listed in Table 8.1.

The first of the three quartic cases is effectively the quartic case arising when γ_1, γ_2, δ are given. The other two quartic cases are equally unamenable to simple solution.

Table 8.1

Given values	Calculation of the other three values involves solving
q r δ q s δ q β δ	a quadratic equation
r s δ r β δ s β δ	a quartic equation

Given q, s, δ

As an example of a quadratic case, consider the one in which we are given the values of q, s, δ and asked to find the value of β.

Subtracting (6) from (5) we obtain

$$pq = rs,$$

and then substituting for r from this in either (5) or (6) we have

$$4\beta^2 s^2 = (p^2 - s^2)(q^2 - s^2); \tag{8}$$

r has now been eliminated. Solving the quadratic (in p) (7) we have

$$p = 2\delta + \sqrt{(q^2 + 4\delta^2)}. \tag{9}$$

(Since p must be positive there is no doubt over which root to take.) We then substitute this value of p in (8) to obtain the desired value of β.

Particular solution

In the notation of the General solution we are given (working throughout in inches)

$$s = 36, \quad q = 48, \quad \delta = 70.$$

Equation (8) of the General solution is now

$$4 \times 36^2\beta^2 = (p^2 - 36^2)(48^2 - 36^2),$$

i.e.

$$36\beta^2 = 7(p^2 - 36^2); \tag{10}$$

and (9) is

$$p = (2 \times 70) + \surd(48^2 + 4 \times 70^2),$$

i.e.

$$p = 288. \tag{11}$$

Substituting from (11) in (10) we obtain

$$36\beta^2 = 7(288^2 - 36^2),$$

i.e.

$$\beta = 126.$$

Hence Ayling Alley is 10 ft 6 in. wide.

Composer's problem

The main question facing the composer is: how hard should the problem be made? In all cases the solver will need to develop for himself the equations (1), (2), (3) (or equivalent ones); the difficulty of his subsequent work will depend on which three unknowns he is given.

For example: if the solver is told that

$$\beta = 126, \quad \gamma_1 = 210, \quad \delta = 70,$$

and asked to find γ_2, then he can immediately calculate successively

$$\text{(from (1))} \quad \alpha_1 = \surd(\gamma_1^2 - \beta^2) = 168,$$

$$\text{(from (3))} \quad \alpha_2 = 1/(1/\delta - 1/\alpha_1) = 120,$$

$$\text{(and from (2))} \quad \gamma_2 = \surd(\alpha_2^2 + \beta^2) = 174.$$

If on the other hand the solver is told that

$$\gamma_1 = 210, \quad \gamma_2 = 174, \quad \delta = 70,$$

and asked to find β, then he is faced with having to solve for β

$$\frac{1}{\surd(210^2 - \beta^2)} + \frac{1}{\surd(174^2 - \beta^2)} = \frac{1}{70},$$

which (unless he banks on the composer's having kindly arranged for an integer solution, and then does some inspired guesswork) must lead him to a most formidable quartic in β^2. That would be

Table 8.2

α	β	γ
4	3	5
3	4	5
12	5	13
8	6	10
24	7	25
6	8	10
15	8	17
12	9	15
40	9	41
24	10	26

too much to ask. So I chose a data set $(\alpha_1 - \alpha_2, \gamma_1 - \gamma_2, \delta)$ that at least required the solution of a quadratic equation, but nothing worse than that.

The other question facing the composer is that of selecting integer values of the data that will lead to an integer solution. This is easily done. We ignore, for the moment, equation (3) of the General solution. We look at the list of 'Pythagorean triplets' (that is, triplets of integers α, β, γ such that $\alpha^2 + \beta^2 = \gamma^2$) listed in Table 8.2 in order of increasing values of β, and take just those for which there are two (or more) entries in the list that have the same value of β (but, of course, different values of α and different values of γ):

β	possible α
8	6, 15
9	12, 40
12	5, 9, 16, 35
15	8, 20, 36, 112
16	12, 30, 63

We now take the pairs of 'possible α' and for each of them calculate δ (we are now renewing our interest in (3) of the General solution) where

$$\frac{1}{\delta} = \frac{1}{\alpha_1} + \frac{1}{\alpha_2}.$$

For example: when $\beta = 8$ we have $\alpha_1 = 6$ and $\alpha_2 = 15$, giving $\delta = 30/7$.

δ having been expressed in its lowest terms, we now 'multiply through' by the denominator of δ; in the example we have chosen this multiplier is 7, and we get

α	β	γ	δ
42	56	70	30
105		119	

as a set of integers satisfying (1), (2), (3) of the General solution.

Extension

The process sketched in the second part of the 'Composer's problem' easily generates integer solutions (as many as we please) of the equations (1), (2), (3) of the General solution; but it also suggests two 'Extension' questions, both of considerable interest, neither of which I can solve.

Parametric solution

If in addition to (1), (2), (3) we impose the further requirement

$$\beta^2 = \alpha_1\alpha_2, \tag{12}$$

then we can obtain — I omit the intermediate working —

$$\left.\begin{aligned}
\beta &= 2kUV(U^2 - V^2)(U^2 + V^2)^2, \\
\delta &= 4kU^2V^2(U^2 - V^2)^2, \\
\alpha_1 &= k(U^2 - V^2)^2(U^2 + V^2)^2, \\
\alpha_2 &= 4kU^2V^2(U^2 + V^2)^2, \\
\gamma_1 &= k(U^2 - V^2)(U^2 + V^2)^3, \\
\gamma_2 &= 2kUV(U^2 + V^2)^3,
\end{aligned}\right\} \tag{13}$$

as a parametric solution: all integer solutions of the four equations (1), (2), (3), (12) are given by (13), where k, U, V ($U > V$; U,V coprime and of opposite parity) take all integer values independently.

But (13) does not, of course, give all the integer solutions of (1), (2), (3): only a very small subset of them.

So the first Extension question is:

Is there a parametric solution (and, if so, what is it) that gives the complete set of integer solutions of the equations (1), (2), (3)?

Least solutions

Defining p, q, r, s as in (4) of the General solution (and assuming for the sake of definiteness that $\alpha_1 > \alpha_2$) we can look for the integer solution of (1), (2), (3) that has the least value of α_1, or of α_2, or of β, or of p, ...

In nine of the ten problems so created, the 'least' solution is one (or more) of the following five solutions

β	δ	α_1	α_2	γ_1	γ_2	p	q	r	s
40	38	399	42	401	58	441	357	459	343
56	30	105	42	119	70	147	63	189	49
63	35	84	60	105	87	144	24	192	18
80	35	84	60	116	100	144	24	216	16
112	14	210	15	238	113	225	195	351	125

where, for example, the underlining of '40' in the first of these five solutions indicates that this solution is the one with least β. Verification of these results is tedious but elementary.

The tenth problem is the one that requires us to find the solution with the least value of s.

At first sight 'min $s = 16$' (from the fourth solution above) is tempting. But further exploration of Pythagorean triplets (see the Composer's problem), while generally producing larger and larger values of s, suddenly yields the solution

β	δ	α_1	α_2	γ_1	γ_2	p	q	r	s
1540	272	561	528	1639	1628	1089	33	3267	11

But is this solution the one with least s? I can not prove that it is, nor can I find a solution with $s < 11$.

So the second Extension question is:

What is the integer solution of (1), (2), (3) that has the least value of s?

9

COUNTERFEIT COINS

Problem I

Tom's Uncle Timothy, a rather senior civil servant, tends to bring his troubles with him when he comes to Ayling for the weekend. On one occasion he confessed to Tom that he was in the devil of a fix: his boss, the Second Lord of the Treasury, had required him to provide the answer to a most delicate problem, and he hadn't a clue how to solve it — he didn't even know where to begin. Shorn of most of Timothy's circumlocutions, the problem was this:

The size, shape, weight, and material of the yet-to-be-announced new 200 p coin are specified with the utmost precision (and are, of course, well known to everybody in the Treasury).

The new coins are already being secretly minted, in considerable numbers, at each of three independent mints.

The Second Lord of the Treasury is satisfied that all the independent mints are following exactly the specifications of size and shape, but he suspects that some — perhaps none, perhaps all — are using a variant material.

Coins made in the variant material will differ slightly in weight from coins made in the correct material; but there is only one possible variant material, so that coins from those mints that use the variant material will all weight the same (though just what they will weigh the Treasury *doesn't* know).

The National Physical Laboratory possesses the only weighing machine on which the Treasury can determine as precisely as necessary the weight of a group of coins — but the NPL will allow their machine to be used only twice.

The Treasury intends to obtain a few coins from each mint, and to carry out the two weighings on the NPL machine — knowing beforehand that from the results they will be able to deduce with certainty which of the mints are using the correct material and which the variant.

The Treasury, being the Treasury, wants the total number of

coins that it has to obtain to be the least necessary for this to be possible.

Timothy has been asked to say what this 'least necessary' total is.

'Tom,' said Uncle Timothy, 'I don't see how, with only two weighings, one can find out which of more than two mints is using the correct material or the variant — however many coins one obtains.' 'I can do that part all right,' said Tom, 'it's the bit about the "least necessary total" that could be the difficulty. However, with only three mints involved, that's fairly easy; it must be . . .'.

What?

Problem II

Tom was rather pleased at having been admitted to the inner workings of the Treasury, and couldn't resist telling Dick all about it. (The writ of the Official Secrets Act doesn't run in Ayling.) Dick wasn't very impressed. 'Lucky for you', he said, 'that there are only three independent mints. Just suppose that there had been four of them — and that everything else was just as your uncle told you† — what would the answer have been then?'

'Tell you tomorrow' said Tom. He did.

What?

† As in Problem I.

Problem III

Telegram from Uncle Timothy to Tom:

Boss's handwriting appalling. What I read as three *mints was really* five *mints. All else the same.*† *Confident you can still do it. Immediate reply essential.*

Telegram from Tom to Uncle Timothy:

Two mints doubles the answer for one mint. Three mints doubles the answer for two mints. Four mints doubles the answer for three mints. Probably five mints doubles the answer for four mints. Working on it.

Second telegram from Tom to Uncle Timothy:

Ignore previous telegram. Answer for five mints is . . .

What?

† The same, that is, as in Problem I.

General solution

The general problem is concerned with N independent mints (rather than with the specific '3', '4', or '5' of the particular problems set).

Let W be the weight of a coin made in the correct material: we know the value of W. Let $W(1 + \varepsilon)$ be the weight of a coin made in the variant material: all we know about ε is that it is not zero.

Let d_r $(r = 0, 1, \ldots, N)$ be 0 or 1 according as $\mathfrak{M}_r$, the rth mint, uses the correct or the variant material: we have that the weight of a coin from $\mathfrak{M}_r$ is $W(1 + d_r\varepsilon)$.

The Treasury's problem is to find out the actual values of the d_r; our problem is to tell the Treasury how to do it.

We are allowed to weigh two groups of coins.

Suppose that in the first weighing we weigh together a group consisting of P_1 coins from $\mathfrak{M}_1$, P_2 coins from $\mathfrak{M}_2$, … , P_N coins from $\mathfrak{M}_N$; and that in the second weighing we weigh together a group consisting of Q_1 coins from $\mathfrak{M}_1$, Q_2 coins from $\mathfrak{M}_2$, … , Q_N coins from $\mathfrak{M}_N$.

The values of the integers $P_1, P_2, \ldots, P_N, Q_1, Q_2, \ldots, Q_N$ are yet to be determined, but are at our disposal.

As a result of these two weighings we know the values of W_1 and W_2, where

$$W_1 = \sum_r W(1 + d_r\varepsilon)P_r,$$

$$W_2 = \sum_r W(1 + d_r\varepsilon)Q_r.$$

Rearranging:

$$(W_1 - W\sum_r P_r)/W = \sum_r d_r P_r\,\varepsilon, \qquad (1)$$

$$(W_2 - W\sum_r Q_r)/W = \sum_r d_r Q_r\,\varepsilon; \qquad (2)$$

where the left-hand sides of (1) and (2) are 'known' (in the sense that they become known as soon as we have allocated specific values to the P_r and the Q_r).

Our first step in allocating values to the P_r and the Q_r is to require that if, for any k, P_k is zero, then Q_k is not zero.

Having imposed this restriction, we can immediately say that if the left-hand sides of both (1) and (2) are zero then all the d_r must be zero: that is, that all the mints are using the correct material.

From now on we consider only the situations in which one (at least) of the left-hand sides of (1) and (2) is non-zero.

From (1), (2) we have

$$\frac{W_1 - W\sum_r P_r}{W_2 - W\sum_r Q_r} = \frac{\sum_r d_r P_r}{\sum_r d_r Q_r} \tag{3}$$

where the left-hand-side of (3) is 'known'.

There are $2^N - 1$ possible combinations of the d_r $(r = 1, 2, \ldots, N)$ apart from the special case, already dealt with, of $d_1 = d_2 = \ldots = d_N = 0$.

If we can select values for P_r and Q_r such that the values of the $2^N - 1$ fractions $\sum d_r P_r / \sum d_r Q_r$ are all different, then the known value of the fraction $(W_1 - W\sum P_r)/(W_2 - W\sum Q_r)$ will determine the values of $d_1, d_2, \ldots, d_N$ uniquely.

Such sets of integers exist: $P_r = 1 \times 2 \times 3 \times \ldots \times r$ and $Q_r = 1$ (for example) meet the requirement† for all values of N.
The number of coins required with such a scheme is, however, unnecessarily large: we are still left with the problem of finding what solution — for a given N — minimizes $\sum \max(P_r, Q_r)$.

Particular solution I

There are three mints to be investigated — $\mathfrak{M}_1, \mathfrak{M}_2, \mathfrak{M}_3$.

First weighing: weigh together one coin from $\mathfrak{M}_2$ and two coins from $\mathfrak{M}_3$ to give

$$W_1 = W\{(1 + d_2\varepsilon) + 2(1 + d_2\varepsilon)\}.$$

Second weighing: weigh together one coin from $\mathfrak{M}_1$ and one coin from $\mathfrak{M}_2$ to give

$$W_2 = W\{(1 + d_1\varepsilon) + (1 + d_2\varepsilon)\}.$$

Then

$$(W_1 - 3W)/W = (d_2 + 2d_3)\,\varepsilon, \tag{1}$$

and

$$(W_2 - 2W)/W = (d_1 + d_2)\,\varepsilon. \tag{2}$$

† The proof is given in the Appendix.

The left-hand-sides of (1) and (2) are known. If both are zero, then we must have that $d_1 = d_2 = d_3 = 0$; and, so, that all the mints are using the correct material.

If one (at least) of the left-hand-sides of (1) and (2) is non-zero, then

$$\frac{W_1 - 3W}{W_2 - 2W} = \frac{d_2 + 2d_3}{d_1 + d_2}, \tag{3}$$

where the left-hand-side of (3) is known.

Since $d_1, d_2, d_3 = 0$ or 1, the seven possible values of the right-hand-side of (3) are as given in Table 9.1.

Table 9.1

d_1	d_2	d_3	$\frac{d_2 + 2d_3}{d_1 + d_2}$
1	0	0	0
0	1	0	1
0	0	1	∞
0	1	1	3
1	0	1	2
1	1	0	½
1	1	1	3/2

Since we know the value of the left-hand-side of (3), and since the values in the right-hand column of the above table are all different, we can immediately deduce from this table the values of d_1, d_2, d_3.

It follows that the Treasury needs a total of four coins: two from one of the mints and one from each of the other two mints.

(That more than three coins are needed is 'almost obvious'; that with four coins the two weighings must be '0 + 1 + 2' and '1 + 1 + 0' is not quite so obvious. Both statements are, however, capable of easy verification.)

Particular solution II

There are four mints to be investigated — $\mathfrak{M}_1, \mathfrak{M}_2, \mathfrak{M}_3, \mathfrak{M}_4$.

There are no solutions when only seven (or less) coins are used, but there are two essentially different solutions when eight coins

Table 9.2

d_1	d_2	d_3	d_4	$\dfrac{d_2 + 2d_3 + 3d_4}{d_1 + 2d_2 + 2d_3}$	$\dfrac{d_2 + d_3 + 4d_4}{2d_1 + d_3 + d_4}$
1	0	0	0	0	0
0	1	0	0	½	∞
0	0	1	0	1	1
0	0	0	1	∞	4
1	1	0	0	⅓	½
1	0	1	0	⅔	⅓
1	0	0	1	3	4/3
0	1	1	0	¾	2
0	1	0	1	2	5
0	0	1	1	5/2	5/2
0	1	1	1	3/2	3
1	0	1	1	5/3	5/4
1	1	0	1	4/3	5/3
1	1	1	0	3/5	⅔
1	1	1	1	6/5	3/2

are used (Table 9.2). (These statements are straightforwardly verifiable, but verification requires a certain amount of effort — which is why Tom wanted a day's grace before giving Dick the answer.)

In one of the two solutions that involve eight coins the Treasury needs three coins from one mint, two coins from each of two other mints, and one coin from the fourth. In the other solution the Treasury requires four coins from one mint, two coins from another mint, and one coin from each of the third and fourth mints.

The argument is essentially that of Particular solution (I). Considering either of the two right-hand columns in Table 9.2, we have that the values in it are all different; and so (knowing the values of W, W_1, W_2) we can immediately deduce the values of d_1, d_2, d_3, d_4.

Particular solution III

There are five mints to be investigated — $\mathfrak{M}_1, \mathfrak{M}_2, \mathfrak{M}_3, \mathfrak{M}_4, \mathfrak{M}_5$. Tom knows that the least number of coins needed in cases involving fewer mints is:

Number of mints	Least number of coins
1	1
2	2
3	4
4	8

and so to begin with he thinks it very likely that, with five mints involved, sixteen coins will be needed. It is only when he carries out the (rather laborious) task of attempting to verify this that — to his surprise — he finds that there is a solution (just one) that needs only fifteen coins.

The Treasury needs to obtain five coins from each of two mints, two coins from each of two other mints, and one coin from the fifth mint.

First weighing: weigh together one coin from $\mathfrak{M}_1$, one coin from $\mathfrak{M}_3$, four coins from $\mathfrak{M}_4$, and five coins from $\mathfrak{M}_5$, to give

$$W_1 = W\{(1 + d_1\varepsilon) + (1 + d_3\varepsilon) + 4(1 + d_4\varepsilon) + 5(1 + d_5\varepsilon)\}.$$

Second weighing: weigh together one coin from $\mathfrak{M}_1$, two coins from $\mathfrak{M}_2$, two coins from $\mathfrak{M}_3$, and five coins from $\mathfrak{M}_4$, to give

$$W_2 = W\{(1 + d_1\varepsilon) + 2(1 + d_2\varepsilon) + 2(1 + d_3\varepsilon) + 5(1 + d_4\varepsilon)\}.$$

Then (ignoring, as before, the special case in which all the mints are using the correct material) we have

$$\frac{W_1 - 11W}{W_2 - 10W} = \frac{d_1 + d_3 + 4d_4 + 5d_5}{d_1 - 2d_2 + 2d_3 + 5d_4} . \qquad (4)$$

The 31 possible values of the right-hand-side of (4) are given in Table 9.3.

Since we know the value of the left-hand-side of (4), and since the values in the 'rhs(4)' column of the above table are all different, we can immediately deduce from the table the values of d_1, d_2, d_3, d_4, d_5.

Table 9.3

d_1	d_2	d_3	d_4	d_5	rhs(3)	d_1	d_2	d_3	d_4	d_5	rhs(4)
						1	1	1	1	1	11/10
1	0	0	0	0	1	0	1	1	1	1	10/9
0	1	0	0	0	0	1	0	1	1	1	11/8
0	0	1	0	0	1/2	1	1	0	1	1	5/4
0	0	0	1	0	4/5	1	1	1	0	1	7/5
0	0	0	0	1	∞	1	1	1	1	0	3/5
1	1	0	0	0	1/3	0	0	1	1	1	10/7
1	0	1	0	0	2/3	0	1	0	1	1	9/7
1	0	0	1	0	5/6	0	1	1	0	1	3/2
1	0	0	0	1	6	0	1	1	1	0	5/9
0	1	1	0	0	1/4	1	0	0	1	1	5/3
0	1	0	1	0	4/7	1	0	1	0	1	7/3
0	1	0	0	1	5/2	1	0	1	1	0	3/4
0	0	1	1	0	5/7	1	1	0	0	1	2
0	0	1	0	1	3	1	1	0	1	0	5/8
0	0	0	1	1	9/5	1	1	1	0	0	2/5

Appendix

Consider the proposition:

Prop (N): d_r ($r = 1,2,\ldots,N$) are, independently, 0 or 1 (with the sole restriction that not all of the d_r are zero). Then the values of the $2^N - 1$ fractions

$$\frac{\sum_{r=1}^{N} d_r r!}{\sum_{r=1}^{N} d_r}$$

are all different.

We now show that Prop($K+1$) is true if Prop(K) is true.

When $N = K+1$ there are $2^{K+1}-1$ fractions, which we separate into two groups: the $2^K - 1$ 'old' fractions that have $d_{K+1} = 0$, and the the 2^K 'new' fractions that have $d_{K+1} = 1$.

Since we are assuming that Prop(K) is true, it follows that the old fractions are all different.

The largest† value of an old fraction is $K!/1$. The smallest† value of a new fraction is

$$\sum_{r=1}^{K+1} r!/(K + 1),$$

† This is not quite immediately obvious, but is easily verified.

which is greater than $K!$. So the values of the new fractions are all greater than the values of all of the old fractions — and, so, different from them.

We still have to show that the values of the new fractions are all different from each other. Suppose that two of them were the same. Then there would exist two old fractions A/B, C/D (with $B, C \leqslant K$) such that

$$\frac{(K+1)! + A}{1+B} = \frac{(K+1)! + C}{1+D} . \qquad (1)$$

Now if $B=D$ we would immediately have $A=C$, and so $A/B = C/D$, which (by Prop (K)) we know to be false; so B and D are unequal. Without loss of generality we can assume $B > D$.

It is readily verified (by considering $\phi(D) - \phi(D-1)$) that, as D increases, the function† of D

$$\phi(D) \equiv (D+1)\sum^{K} r! - (D+2)\sum^{D} r!$$

increases from $\phi(1)$ to

$$\phi(K-2) = \phi(K-1) = (K+1)! - \sum^{K} r!.$$

So, D being an integer, we certainly have that, for $1 \leqslant D \leqslant K-1$,

$$(D+1)\sum^{K} r! - (D+2)\sum^{D} r! < (K+1)! \qquad (2)$$

Hence, since $B - D \geqslant 1$,

$$(D+1)\{\sum^{K} r! - \sum^{D} r!\} < (B-D)\{\sum^{D} r! + (K+1)!\}, \qquad (3)$$

which we rearrange as

$$\frac{(K+1)! + \sum^{K} r!}{1+B} < \frac{(K+1)! + \sum^{D} r!}{1+D} . \qquad (4)$$

† Why now introduce this particular function? Is it an inspired guess? No: what happened in practice was what I said: 'I want to prove (5); I can prove (5) if I can prove (4); I can prove (4) if I can prove (3); I can prove (3) if I can prove (2)'. Having proved (2) I then present the whole proof in the reverse order: (2) → (3) → (4) → (5). More logical but more mysterious, as is often the case with mathematical proofs.

Now $A \leqslant \sum^{K} r!$, and $C \geqslant \sum^{D} r!$, so

$$\frac{(K+1)! + A}{1 + B} < \frac{(K+1)! + C}{1 + D}, \tag{5}$$

contrary to (1). It follows that the new fractions are all different from each other.

Consequently Prop(K+1) is true if Prop(K) is true; and so, since Prop(1) is true, we have that Prop(N) is true for all $N \geqslant 1$.

Extension

We can show (as in the Appendix) that *a* solution exists for any value of N; but obtaining the *least* solution is a different matter. I have found this problem peculiarly intractable.

Tom (that is, I) hazarded a guess that the least solution would involve a total of 2^{N-1} coins for N mints — and then found that for five mints only fifteen (rather than sixteen) coins were needed.

On the other hand, I have been unable to show that 2^{N-1} coins are sufficient for all values of N.

So the extension questions are:

1. *Are 2^{N-1} coins always sufficient, for general values of N?*
2. *What is the least number of coins needed, for general values of N?*

10

WRAPPING A PARCEL

Problem

The Beeling Biscuit Bakery packs its products in rectangular boxes 7″ × 8″ × 11″.

Dick has the wrapping contract with the Bakery; to fulfil it he has bought a large quantity of wrapping paper in rectangular sheets 18″ × 30″.

He wraps each box in a single sheet of paper in the normal way. Admittedly there are places where the edges of the paper only just meet round the box, without any overlap; but sticky tape takes care of that, and the terms of his contract require only that the surface of the box should be completely covered with a single sheet of wrapping paper (without stretching or tearing it).

Yesterday the Beeling Biscuit Bakery told Dick that in order to accommodate the Bigger Beeling Biscuit they were going to change the size — though not the shape — of their boxes. The linear dimensions are to be increased by one part in *B* — but it has not yet been decided what *B* will be.

Dick was despondent; but today (after having turned and twisted a bit†) he realizes that he will still be able to use his existing stock of wrapping paper, and still meet the terms of his contract, provided that *B* is not less than . . .

What?

† This is a clue!

General solution

Let $\mathscr{C}$ be a cuboid — a box, for example — with edge-lengths† L, M, S ($L \geqslant M \geqslant S$). Let $\mathscr{R}$ be a rectangular sheet of paper with sides U, V.

Normal‡ Wrapping

Setting out to wrap a box, most people tend to put it down on the paper with its short edges vertical, and its other edges parallel to the sides of the paper (see Diagram 10.1a). Then (Diagram 10.1b,c), for the wrapping completely to cover the box, without stretching or tearing the paper, the sides U,V of the paper need to be (at least)

$$2(S + M), \quad S + L;$$

or

$$2(S + L), \quad S + M.$$

In either case the area of the paper needs to be (at least)

$$\mathscr{A}_N = 2(S + M)(S + L). \tag{1}$$

I call this 'normal wrapping of the first kind'.

Another way of wrapping the box is to put it down on the paper with either its medium or its long edges vertical, and its other edges parallel to the sides of the paper (Diagram 10.2a). Then (Diagram 10.2b,c) the sides U,V of the paper need to be (at least)

$$2(S + M), \quad S + L \qquad \text{(medium edges vertical),}$$

or

$$2(S + L), \quad S + M \qquad \text{(long edges vertical).}$$

In either case the area of the paper needs to be (at least) $\mathscr{A}_N$ (as defined in (1)). I call this 'normal wrapping of the second kind'.

Clearly 'normal wrapping of the second kind' provides no improvement over 'normal wrapping of the first kind' — which is probably why it is so rarely used, since (as between the two) the 'first kind' is the easier to perform.

† 'Long, Medium, Short'.
‡ 'Normal' in all of the senses 'rectangular', 'perpendicular', 'regular', 'usual'.

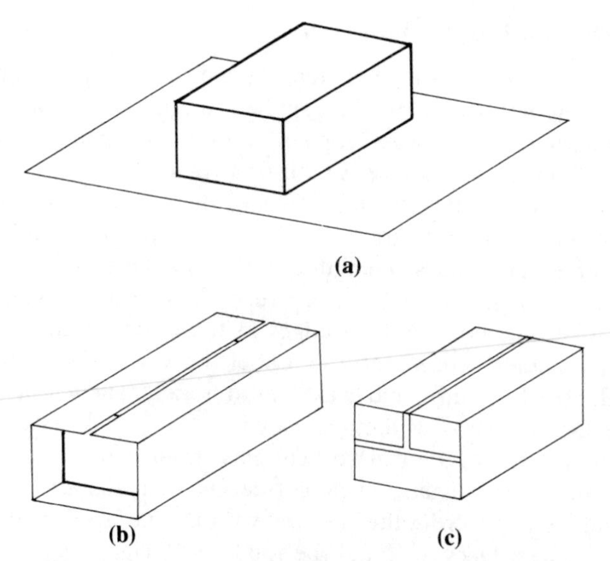

Diagram 10.1. Normal wrapping of the first kind

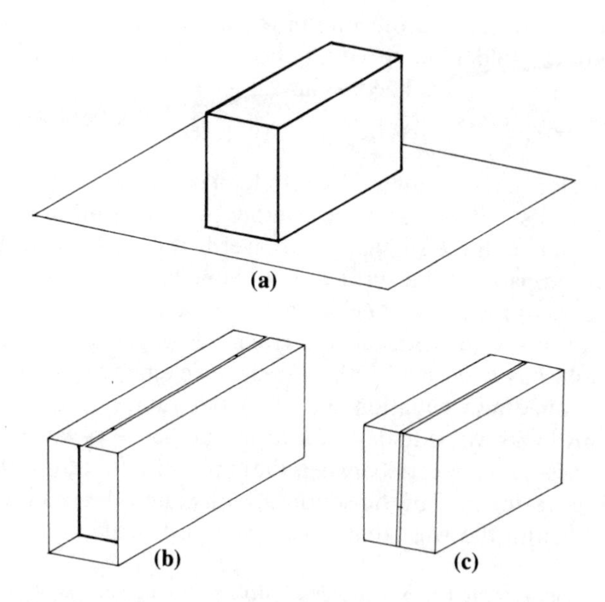

Diagram 10.2. Normal wrapping of the second kind

Skew-rectangular Wrapping

There is, however, a further step that we can take which gives an immediate improvement when applied to normal wrapping of the second kind (but not when applied to normal wrapping of the first kind). Simply, it is to give $\mathscr{C}$ a bit of a twist.

Let us start again with the initial position for normal wrapping of the second kind (Diagram 10.2(a)), with the medium† edges vertical. $\mathscr{R}$ has its sides of length U in the east–west direction. $\mathscr{C}$ has the centre of its base at the centre of $\mathscr{R}$; its short edges in the east–west direction; its long edges in the north–south direction; and its medium edges vertical. Label the vertical faces of $\mathscr{C}$ as 'north', 'east', 'south', and 'west' appropriately (and let them keep those labels even after the next step).

Now rotate $\mathscr{C}$ (about a vertical axis through the centre of its base) through a smallish angle ϕ (see Diagram 10.3a.)

Fold the paper from the east and west up and over to cover the east and west faces of $\mathscr{C}$ and the top of $\mathscr{C}$ (Diagram 10.3(b)); we need to have

$$U \geqslant 2(S + M) \cos \phi. \tag{2}$$

There is now paper projecting in the planes of the east and west faces of $\mathscr{C}$; fold it in to cover the north and south faces of $\mathscr{C}$ (Diagram 10.3c). We need to have

$$V \geqslant (S + L) \cos \phi. \tag{3}$$

It looks for the moment as if we could make the area of $\mathscr{R}$ (namely $U \times V$) as small as we please, simply by increasing the angle ϕ; but, of course, there's a snag: if we overdo it we will create gaps — uncovered parts of the surface of $\mathscr{C}$. How far can we go?

One way in which to decide that is to take axes parallel to the edges of $\mathscr{C}$ and to work out the equations of the sides of $\mathscr{R}$ after each fold has been made; the algebra is elementary (there's not even a quadratic equation to solve) but rather lengthy, and it certainly looks worse than it is. It turns out that — as we increase ϕ — the first gap appears between the top of the fold from the east that is covering part of the south face of $\mathscr{C}$, and the south edge of the fold from the west that is covering part of the top of $\mathscr{C}$. To

† The improvement that we are going to develop also occurs if we start with the long edges vertical, but then it is not so great.

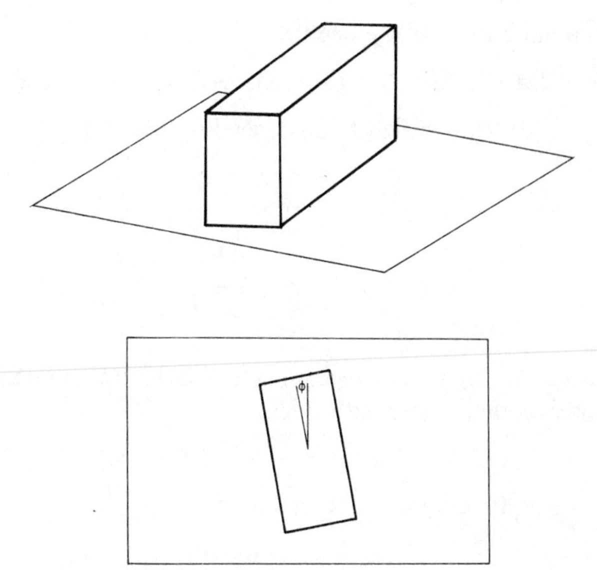

Diagram 10.3(a). Skew-rectangular wrapping: start

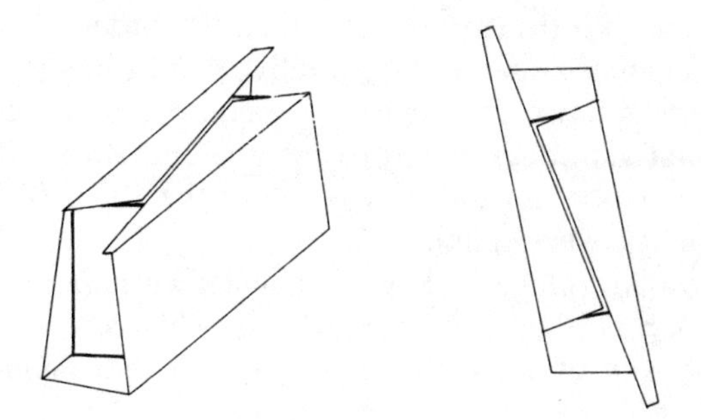

Diagram 10.3(b). Skew-rectangular wrapping: after first folds

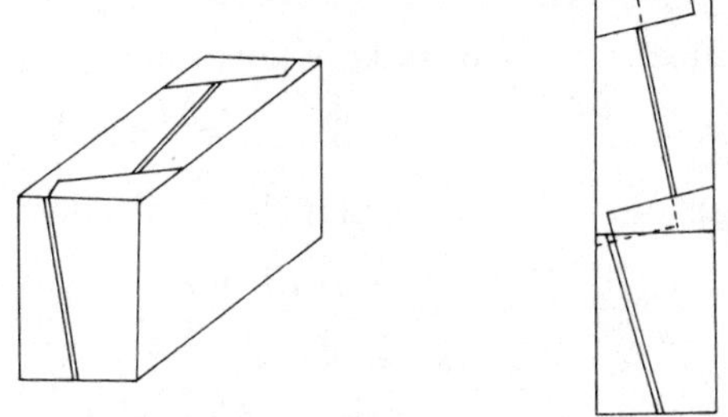

Diagram 10.3(c). Skew-rectangular wrapping: after final folds

avoid creating this gap we need

$$(3S + 2M + L)\sin\phi + (S + 2M + L)\cos\phi \leqslant U + V. \quad (4)$$

From (2), (3), (4) we find that we need to have

$$\phi \leqslant \alpha$$

where

$$\tan\alpha = \frac{2S}{3S + 2M + L}.$$

We have, then, that $\mathscr{C}$ can be completely covered by a rectangular sheet of wrapping paper with sides

$$2(S + M)\cos\alpha, \quad (S + L)\cos\alpha,$$

where α is defined by (5). Its area is

$$\mathscr{A}_R = \mathscr{A}_N \cos^2\alpha,$$

where $\mathscr{A}_N$ is defined by (1). I call this 'skew-rectangular wrapping'.

It may be worth emphasizing that, in order to carry out skew-rectangular wrapping successfully, $\mathscr{C}$ must be put on $\mathscr{R}$ with its short edges horizontal and its medium edges vertical; that $\mathscr{C}$ must be rotated through an angle (not greater than) α; and that the first folds must be 'up and over' to cover the '$M \times L$' faces of $\mathscr{C}$ and then meet over the top of $\mathscr{C}$.

The greatest advantage of skew-rectangular wrapping over normal wrapping occurs when $\mathscr{C}$ is a cube (that is, when $S = M = L$): here $\tan\alpha = 1/3$, and so $\mathscr{A}_R = 9/10\,\mathscr{A}_N$ — a ten per cent saving.

Particular solution

The original Beeling Biscuit Bakery boxes have edges

$$S = 7'', \quad M = 8'', \quad L = 11'';$$

and Dick wraps them in rectangular sheets of sides

$$U = 30'' = 2(S + M),$$

$$V = 18'' = S + L.$$

From the General solution we know that he could in fact wrap them in rectangular sheets of sides

$$30'' \cos \alpha, \qquad 18'' \cos \alpha,$$

where

$$\tan \alpha = \frac{2 \times 7''}{(3 \times 7'') + (2 \times 8'') + 11''}$$

$$= \frac{7}{24}.$$

from which

$$\cos \alpha = \frac{24}{25}.$$

So if the new Beeling Biscuit Bakery boxes are to have edges

$$7'' \times (1 + 1/B),$$

$$8'' \times (1 + 1/B),$$

$$11'' \times (1 + 1/B),$$

Dick can wrap them successfully, using his $30'' \times 18''$ wrapping paper, provided that

$$1 + 1/B \leqslant \sec \alpha;$$

that is, provided that

$$B \geqslant 24.$$

Composer's problem

In the days when bread was bread, and one sliced one's own, one bought a loaf from the man who baked it; and the baker's assistant wrapped it in a sheet of tissue-paper before one's eyes. Most things to be wrapped are most easily put down on the paper with the short edges vertical; but a loaf (that is, a loaf that is loaf shaped, as loaves used to be) is most easily put down with the medium edges vertical. So the assistant did just that — and invariably put it down 'skew'. Ten years old, I idly wondered why; and forgot about it. Thirty years later I remembered the problem, and solved it (for the cuboid — the 'rectangular box' — which is of course much easier to deal with than a loaf shape). Obviously the problem and its solution are not 'new': bakers' assistants solved it and took advantage of skew-rectangular wrapping forty years ago — and probably far longer ago than that.

Setting a particular problem on skew-rectangular wrapping is easy: even if one imposes on oneself the requirement that all the edges of the parcel are to be integers and the value of $\cos\alpha$ (in the General solution) is to be rational — as I did — all one needs to do is to select integer values of L, M, S ($L \geqslant M \geqslant S$) such that $(2S)^2 + (3S + 2M + L)^2$ is itself the square of an integer. I just chose the least set (L,M,S) that meets that requirement.

Extension

The use of 'normal wrapping' implicitly imposes two restrictions — that the wrapping paper is rectangular, and that the box is put down on the paper with edges parallel to the sides of the paper.

'Skew-rectangular wrapping' recognizes the second of those restrictions, and removes it: an improvement results — for any given box, the minimum area of the necessary wrapping paper is reduced.

But what happens if we remove (also) the first restriction?

I have taken this question a little way — by allowing the wrapper to be a parallelogram rather than confining it to be a rectangle. In that case the algebra (though still basically elementary) becomes very turgid — certainly not fit to be reproduced here. The results are:

'*Long box*': If

$$3S + 2M - L < 0$$

then the relaxation does give an improvement — but a slight one: the wrapping area required is reduced by — at most — one part in 274.

'*Flat box*': If

$$S - 2M + L < 0$$

then the relaxation again gives an improvement — but a very slight one: the wrapping area required is reduced by — at most — one part in 1229.

'*Middling box*': If both

$$3S + 2M - L \geqslant 0$$

and

$$S - 2M + L \geqslant 0$$

then the relaxation gives no improvement at all: the most economical parallelogram is a rectangle.

If one imposes no restrictions at all on the shape of the wrapper, then the area of the minimum wrapper becomes, simply, the surface area of the box: which is very uninteresting.
But, in between, there are two questions that are of interest — and which I have not solved.

1. *What (in terms of L, M, S — the edge-lengths of the box) is the minimum area of the wrapper, subject only to the restriction that the wrapper is convex?*
2. *What (in terms of L, M, S — the edge-lengths of the box) is the minimum area of the wrapper, subject only to the restriction that the wrapper is a plane-filler — that is, that we can cut as many wrappers as we please out of an indefinitely large sheet of paper without wasting any of it?*

11

SHEEP-DOG TRIALS

Problem†

Ceiling shepherds view orthodox sheep-dog trials with some disdain; their own, they consider, are much more practical (they take up less room; they involve only one sheep for each trial; and the sheep are competitors, too).

The Ceiling Sheep Circle is simply a circular piece of land surrounded by a low wall (so low as to present no deterrent at all to any sheep).

There are just two rules in a Ceiling Sheep-Dog Trial:

I. No dog may ever enter the Ceiling Sheep Circle.

II. A trial starts with a sheep being released at the centre of the Circle; it stops when the sheep escapes from the Circle. The score is the elapsed time. (The longer the time, the better the dog's score; the shorter the time, the better the sheep's score.)

Now just as Ceiling sheep-dogs are trained never to enter the Circle (and are very intelligent), so also Ceiling sheep are trained to try to escape from the Circle (and also are very intelligent). There is no question of the sheep just standing there grazing and the dog just lying there looking at it; that sheep is as determined to win the inter-sheep competition as the dog is to win the inter-dog one.

This year there came up against each other a sheep that could run at only 5 m.p.h. and a dog that could run at U m.p.h. (U being a whole number). It was very exciting — in fact‡ there was a

† Solving this problem involves — unfortunately — some knowledge of trigonometry; and (even more unfortunately) access to tables of some trigonometric functions, or to a calculator that can cope with them.

‡ If one can bring oneself to think of sheep and dogs as points, then the point-sheep escaped — but it was still a close-run thing!

head-on collision: the sheep had got its nose out of the Circle *very* shortly before the dog arrived, and felt that that counted as an escape. They're still arguing about it in Ceiling.

What was the value of U?

General solution

1. A sheep S, capable of speed V, is at the centre O of a circle $\mathscr{C}$ of radius R. A dog D, capable of speed U, is just outside $\mathscr{C}$, but is not permitted to enter $\mathscr{C}$. Can the dog prevent the sheep's escaping from $\mathscr{C}$?

2. If $U \leqslant V$ the answer is obvious. So from now on we assume that $U > V$. Let†

$$U/V = \sec \alpha.$$

3. The simplest way of presenting the solution is, I think, to ascribe a specific strategy to the sheep (without asserting that it is the best that the sheep could adopt); then to calculate α^*, a number for which we can say 'if the sheep adopts the ascribed strategy, and $\alpha < \alpha^*$, then the sheep will escape, whatever the dog does'; and then separately to show that if $\alpha \geqslant \alpha^*$ the dog can prevent the sheep's escaping, whatever the sheep does.

4. We define $\mathscr{B}$ to be the circle with centre O and radius $R \cos \alpha$.

(The significance of this circle is that, when the sheep is inside $\mathscr{B}$, it can run so as to maintain — or get back to — a situation in which O is between it and the dog, whatever the dog does; whereas while the sheep is outside $\mathscr{B}$ the dog can reduce $\hat{\text{DOS}}$, whatever the sheep does.)

5. The strategy that we ascribe to the sheep is:

(i) Always run at speed V.

(ii) Always move so as to increase the distance SO.

(iii) While within $\mathscr{B}$: keep O between S and D.

(iv) While on or outside $\mathscr{B}$: there are two directions from S that satisfy (ii) and that are tangential to $\mathscr{B}$: they meet $\mathscr{C}$ in (say) T′, T″: move towards T′ or T″ according as DT′ is greater or less than DT″ (and if DT′ = DT″ select either direction indifferently).

6.1. Following this strategy the sheep first runs, keeping O between itself and the dog, until it reaches $\mathscr{B}$; at Q, say.

† Why 'sec α' rather than something simple, like 'k'? Mainly because we find — later — that the angle α comes naturally and frequently into the analysis, and using it from the start simplifies the presentation.

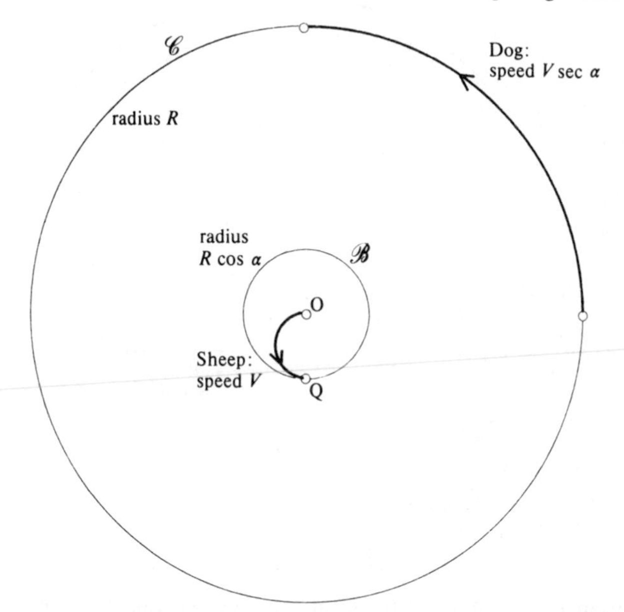

Diagram 11.1. First phase of the sheep's path: if the dog runs round $\mathscr{C}$ at speed V secα, the sheep runs at speed V keeping O between itself and the dog: the sheep's path is a semicircle of radius $\frac{1}{2}R$ cos α.

(This phase takes the sheep time $\pi(R \cos \alpha)/2V$ at most. It needs $\pi(R \cos \alpha)/2V$ if the dog runs round $\mathscr{C}$ at speed U. Diagram 11.1 illustrates the situation in which the dog runs round $\mathscr{C}$ at speed U always in the same direction; the sheep's path is then a semicircle of radius $\frac{1}{2}R \cos \alpha$.)

6.2. The sheep then leaves $\mathscr{B}$ tangentially. If, as in Diagram 11.1, the dog has been running round $\mathscr{C}$, and continues to do so in the same direction, then the sheep's path from Q is simply QT′, the tangent to $\mathscr{B}$ at Q, and its point of arrival at $\mathscr{C}$ is T′, as in Diagram 11.2a.

(This is in fact the best course of action for the dog to adopt — but it may decide not to adopt it, which is why the sheep still needs to keep an eye on the dog's movements.)

If the dog turns and runs in the opposite direction round $\mathscr{C}$, the time may come when it crosses the (extended) line SO, in which

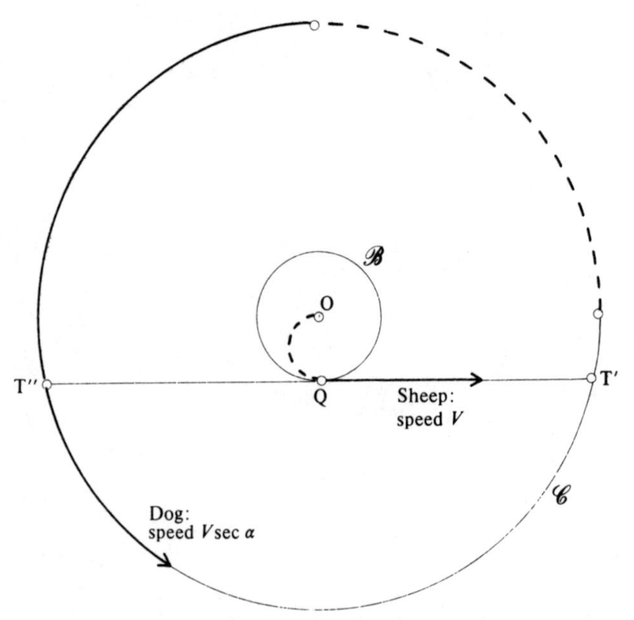

Diagram 11.2(a). Second phase of the sheep's path: if the dog goes on running round $\mathscr{C}$ at speed V sec α, the sheep runs at speed V along the tangent to $\mathscr{B}$ at O.

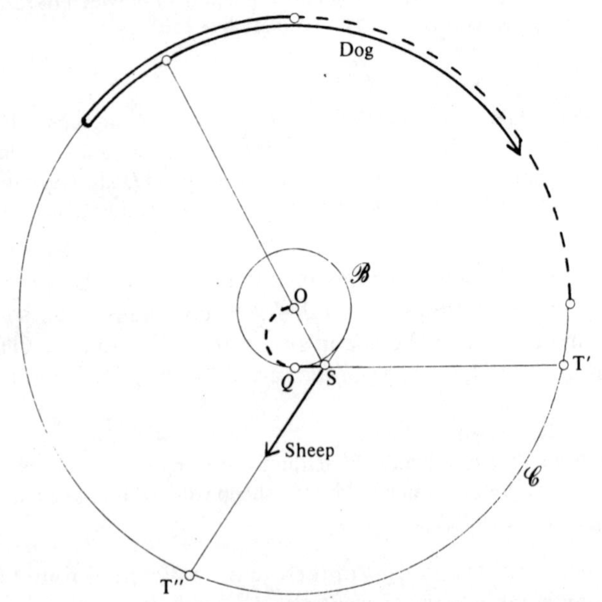

Diagram 11.2(b). Second phase of the sheep's path: if the dog (having turned and run the other way round $\mathscr{C}$) crosses SO (extended), the sheep changes course to the other tangent to $\mathscr{B}$.

case the sheep then changes course in accordance with point (iv) of its strategy (as in Diagram 11.2(b)).

(There may be several such changes of course. It is perhaps worth emphasizing that it is not the dog's reversal of direction that causes the sheep to change course; the sheep changes course only if the dog crosses the extension of SO.)

6.3. On the final occasion on which the dog is on the extension of SO let the sheep's distance from O be $R \cos \alpha \sec \beta$. We have that

$$0 \leqslant \beta < \alpha < \pi/2 \tag{1}$$

The sheep's subsequent course is along ST′, and the dog's is from D towards T′ by way of T″ (see Diagram 11.3).

Now

$$\mathrm{ST}' = R(\sin \alpha - \cos \alpha \tan \beta),$$

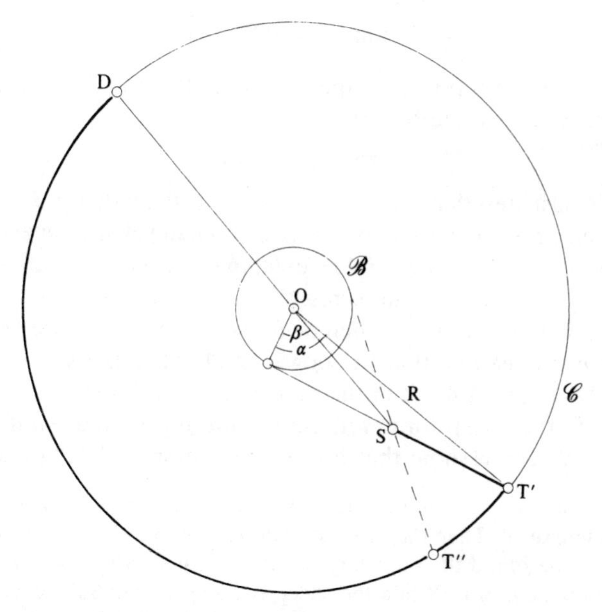

Diagram 11.3. On the last occasion on which the dog crosses SO (extended), the sheep is at S distant $R \cos \alpha \sec \beta$ from O. The sheep runs from S to T′; the dog runs toward T′ by way of T″.

and so the sheep covers the distance ST′ in time t_S, where

$$Vt_S = R(\sin\alpha - \cos\alpha \tan\beta). \tag{2}$$

The dog covers the major arc DT′ of $\mathscr{C}$ in time t_D, where

$$Vt_D = R(\pi + \alpha - \beta)\cos\alpha. \tag{3}$$

By (2), (3) we have

$$V(t_D - t_S) = R(\pi - \tan\alpha + \alpha + \tan\beta - \beta)\cos\alpha. \tag{4}$$

We digress for a moment to note that $\tan\theta - \theta$ $(0 \leqslant \theta < \tfrac{1}{2}\pi)$ is an increasing function of θ (as is easily verified). Using this fact first with $\theta = \beta$ and then with $\theta = \alpha$ we have from (1), (4)

$$V(t_D - t_S) \geqslant R(\pi - \tan\alpha + \alpha)\cos\alpha, \tag{5}$$

and then from (5) that

$$t_D - t_S > 0$$

if $\alpha < \alpha^*$, where α^* is defined by

$$\tan\alpha^* - \alpha^* = \pi \tag{6}$$

That is, the sheep will escape, whatever the dog does, if $\alpha < \alpha^*$, where α^* is defined by (6).

7. The last step that we need to take is to show that if $\alpha \geqslant \alpha^*$ the dog can prevent the sheep's escaping, whatever the sheep does.

The sheep may leave and re-enter $\mathscr{B}$ several times, but — if it is to escape from $\mathscr{C}$ — there must be a final occasion on which it leaves $\mathscr{B}$. Let its position then be S′, and let S′O extended cut $\mathscr{C}$ in D′; we can assume that the dog is at D′ when the sheep is at S′ (See Diagram 11.4.). Let the tangent to $\mathscr{B}$ at S′ meet $\mathscr{C}$ in T′, T″. From S′ the sheep runs, without re-entering $\mathscr{B}$, to a point, P say, on $\mathscr{C}$. We can assume that P is on the minor arc T′T″ of $\mathscr{C}$.

(This may not be immediately obvious. But suppose that P were elsewhere on $\mathscr{C}$. Then the shortest path from S′ to P that did not re-enter $\mathscr{B}$ would be round the boundary of $\mathscr{B}$ to S″, where S″P is tangential to $\mathscr{B}$ and then along S″P. While the sheep covered the arc S′S″ of $\mathscr{B}$ the dog could cover the corresponding arc D′D″ of $\mathscr{C}$: the situation with the sheep at S″ and the dog at D″ would be rotationally identical with that with the sheep at S′, the dog at D′, and P at T′.)

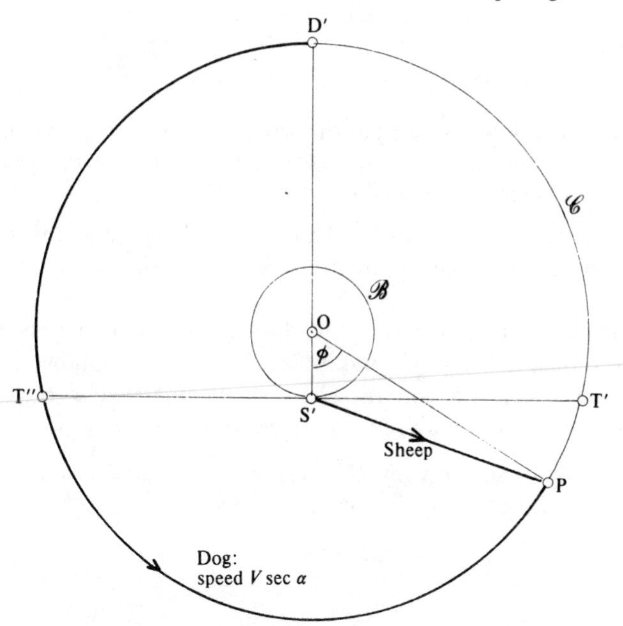

Diagram 11.4. Second phase of the sheep's path: if the sheep runs from S′ towards P, the dog runs round $\mathscr{C}$ at speed V sec α.

Let $S'\widehat{O}P = \phi$; we have that

$$-\alpha \leqslant \phi \leqslant \alpha. \tag{9}$$

To get from S′ to P the sheep needs time t_S, where

$$Vt_S = R\surd(1 - \cos\alpha\cos\phi + \cos^2\alpha). \tag{10}$$

The dog, starting from D′ and going round the major arc of $\mathscr{C}$, can reach P in time t_D, where

$$Vt_D = R(\pi + \phi)\cos\alpha. \tag{11}$$

From (10), (11) we have

$$(V/R)(t_S - t_D)\sec\alpha =$$

$$g(\phi) \equiv \surd(\sec^2\alpha - 2\sec\alpha\cos\phi + 1) - \pi - \phi, \tag{12}$$

which is a a decreasing function of ϕ in the range (9).

Hence

$$(V/R)\,(t_S - t_D)\sec\alpha \geqslant \tan\alpha - \pi - \alpha. \qquad (13)$$

(It may be worth pointing out that $g(\phi)$ does *not* have a minimum at $\phi = \alpha$, even though its first derivative is zero there. The argument from (12) to (13) depends essentially on (9).)

It follows from (13) that, whatever the sheep does, the dog can prevent the sheep's escaping if $\alpha \geqslant \alpha^*$, where α^* is defined by (6).

8. In Section 6 on page 88 we have shown that the sheep can always escape if $\alpha < \alpha^*$, and in Section 7 we have shown that the dog can always prevent the sheep's escaping if $\alpha \geqslant \alpha^*$. So the answer to the original question in Section 1 is:

The dog can prevent the sheep's escaping if and only if $U/V \geqslant \sec\alpha^*$, where α^* is defined by

$$\tan\alpha^* - \alpha^* = \pi.$$

Particular solution

The sheep runs at 5 m.p.h., and the dog at U m.p.h. So we have from the General solution that the sheep can escape from the Circle if and only if $U < 5\sec\alpha^*$, where α^* is defined by

$$\tan\alpha^* - \alpha^* = \pi.$$

You will have to take my word for it (or your pocket calculator's) that

$$4{\cdot}6 < \sec\alpha^* < 4{\cdot}8.$$

So the sheep can escape if $U \leqslant 23$, but not if $U \geqslant 24$.
So

$$U = 23.$$

(If we take the sheep's speed as 5 m.p.h. and the dog's as 23 m.p.h., and assume that the Ceiling Sheep Circle has a radius of 100 yards — which is a fair sized circle — then the sheep gets about 2½ inches out of the Circle by the time the dog arrives: that is, if it is a point-sheep and the dog is a point-dog.)

Composer's problem

I first met this problem about thirty years ago (where, I do not remember); solved it; and forgot it. It was at that time posed in the form of a girl swimming in a circular lake, with an eager young man — who couldn't swim — on the bank of the lake: could the girl get away from the man? I felt that that formulation was not entirely satisfactory, since it was likely that the man could run faster than the girl could — especially a girl who had been doing some energetic swimming — and so merely getting out of the lake successfully was not going to help her in the long run.

Ten years ago I looked at the problem again, recasting the participants so that the swimmer was a celibate man and the eagerness was all on the part of a girl on the bank. I also generalized: allowing the girl to be able to swim — and swim faster than the man could. That problem kept me busy for most of my spare time over several months: the result was an addition to my files of forty pages of (elementary) trigonometry and diagrams that nobody (myself included) would ever want to look at again.

So, when I went through my notes looking for material for this book, 'The case of the celibate swimmer' did not get on to any list (never mind the short list), for two reasons: first, the problem was 'well known' (and of long standing); second, it involved trigonometry, approximation, and a messy solution (like it or not, $\tan \alpha^* - \alpha^* = \pi$ is a messy definition of α^*).

What made me change my mind? Watching 'One man and his dog' on television. The real point of setting this problem here is to provide an excuse for mentioning the much more difficult problem I pose in the Extension section: one that should keep the theorists (mathematical, bucolic, and perhaps even military) busy for a considerable period. (There might even be a computer game in it.)

Extension

Ceiling shepherds may view orthodox sheep-dog trials with disdain — but they still watch 'One man and his dog' on television. They have been particularly interested in the Pairs competitions, and have decided to have a Pairs event at Ceiling next year. The rules will be just as for the traditional Singles event, but there will be *two* dogs (starting from diametrically opposite positions) just outside the Circle when the sheep is released at the centre of the Circle.

Every dog and every sheep in Ceiling knows the critical ratio sec α^* ($\simeq 4{\cdot}6033$) for Singles events: but they are all having sleepless nights trying to work out what (even on the simplifying assumption that the two dogs are capable of the same top speed) the critical ratio is for a Pairs event. And what the sheep's best strategy is. And what the dogs' best strategy is.

I have been unable to help them. They (and I) would very much like to know the answer.

ANSWERS

Ladder–box	17 ft
Meta-ladder–box	10 ft
Complete quadrilateral	40 square miles
Bowling averages (I)	7
Bowling averages (II)	29
Centre-point	6
Counting Sheep	210 square yards
Transport (I)	1.52 p.m.
Transport (II)	1.51 p.m.
Transport (III)	1.32 p.m.
Alley ladder	10 ft 6 in.
Counterfeit coins (I)	4
Counterfeit coins (II)	8
Counterfeit coins (III)	15
Wrapping a parcel	24
Sheep-dog trials	23 m.p.h.